A Groundbreaking Integration of Quantum Physics,
Cellular Biology, and the Wisdom of the Mind and Heart

ATOMIC GLOW

THE FORGOTTEN LIFE CODE ON EARTH

ALISON DAVIS

First published by Ultimate World Publishing 2025
Copyright © 2025 Alison Davis

ISBN

Paperback: 978-1-923583-38-2
Ebook: 978-1-923583-39-9

Alison Davis has asserted her rights under the Copyright, Designs and Patents Act 1988 to be identified as the author of this work. The information in this book is intended for general knowledge and inspiration only. It is not a substitute for professional medical advice, diagnosis, or treatment. Always consult a qualified doctor or health practitioner regarding any medical concerns or before making changes to your health practices. The author and publisher accept no responsibility for any adverse outcomes resulting from the use or misuse of the information contained in this book. Readers are encouraged to take full responsibility for their own health choices and wellbeing.

All rights reserved. No part of this publication may be reproduced, stored in or introduced into a retrieval system, or transmitted in any form, or by any means (electronic, mechanical, photocopying, recording or otherwise) without the prior written permission of the author. For permission requests, please contact the publisher.

Cover design: Ultimate World Publishing
Layout and typesetting: Ultimate World Publishing
Editor: Rebecca Low

Ultimate World Publishing
Diamond Creek,
Victoria Australia 3089
www.writeabook.com.au

Acknowledgments

The wisdom in this book would not exist without the extraordinary mentors who illuminated my path over the past decade of transformation. Some I was blessed to meet in person, others I only met through their words and teachings, yet each of them became a lighthouse at the very moment I needed guidance, shining truth, awakening courage, and helping me take the next step. Their light led me to discover the lighthouse within myself, so I, too, may stand as a beacon for others when their moment of need arrives.

After reading over 500 books and listening to countless podcasts and lectures, I am forever grateful for the brilliant minds and beautiful souls who touched my life. Shawn Achor showed me that happiness was possible and sparked the journey. Dr. Joe Dispenza awakened me to the power of the mind and heart. Eckhart Tolle revealed the truth of who I really am. Neale Donald Walsch and Dr. Wayne Dyer gave me a taste of the divine power we each hold. Michael Singer offered the art of surrender. Dr. Bruce Lipton opened my eyes with his groundbreaking work in epigenetics. And most profoundly, Dr. Jack Kruse, whose revolutionary insights into quantum biology and physics reshaped not only how I see life, but how I live it.

ATOMIC GLOW

I also honour the many visionaries who deepened my path of self-mastery and spiritual expansion, Gary Zukav, David Hawkins, Dr. Jill Bolte Taylor, Byron Katie, and many others. To all of you: thank you for reminding me who I truly am.

To my beautiful family. This book carries your fingerprints on every page. Without your love and inspiration, it may never have been written. You came into my life not just to walk beside me, but to remind me why I am here: to grow, to remember, and to share my gift with the world. This journey is as much yours as it is mine.

And to you, my dear reader. I honour your thirst for wisdom and your courage to keep stepping forward. I acknowledge your willingness to rise even when life tests you, and your strength in being the rock for your family. I see your openness to growth, your light that touches others in ways you may never know, and your quiet yet powerful role in shaping a more conscious, compassionate world. Thank you for joining me, choosing to glow, and helping humanity itself evolve.

DEDICATION

To my incredible sons, Alexander and Xavier.

You are my teachers, my mirrors, my greatest source of joy. You've shown me what unconditional love looks like, what real freedom feels like, and what resilience and purpose truly mean. You are, and always will be, my why.

Contents

Acknowledgments — iii
Dedication — v
Introduction — 1
Chapter 1: The Early Script of Life — 5
Chapter 2: You Were Born Lucky — 15
Chapter 3: The Motherhood Fall — 25
Chapter 4: Game of Life in the Mind — 35
Chapter 5: Deep Inside Our Universal Mind — 47
Chapter 6: The Power of a Broken Heart — 61
Chapter 7: The Activator of the Atomic Universe — 73
Chapter 8: The Evolution of Life on Earth — 87
Chapter 9: The Engines of Human Life — 99
Chapter 10: Optimal Inputs: Fuels for the Perfect Engine — 109
Chapter 11: Vital Outputs: The Power You Become — 125
Chapter 12: Modern Disruptors of Your Glow — 143
Chapter 13: Nature-Inspired Modern Living — 161
Chapter 14: The Innate Tuning Forks of Reality — 173
Chapter 15: The Art of Atomic Glow — 189
Afterword: Our Future—The Era of the Heart — 201
About the Author — 203
Appendix A: Mitochondrial Inputs & Outputs — 205
References — 209

Introduction

In the eyes of many people living in the modern world, life can feel overwhelming. The constant stream of uncertainties, global conflicts, societal pressures, and personal struggles can leave even the strongest among us feeling lost, anxious, confused, even scared and, at times, hopeless. It's easy to believe that the world is spiralling out of control and we're just trying to survive within it.

And yet, there are others, perhaps fewer in number, who see life through a completely different lens. They experience the same world, the same ups and downs, similar triumphs and challenges, but they meet it with a grounded sense of purpose. They don't deny adversity or uncertainty; they don't shrink from it either. Instead, they face it with curiosity and embrace it with courage. They see setbacks as setups for comebacks, pain as a path to wisdom, and the unknown as a gateway to possibility. Even in the midst of chaos, they feel a quiet confidence, a deep sense of freedom, and a trust in the unfolding of their lives.

What separates these two experiences?

It's not the circumstances. Adversities and uncertainties are not new; they've existed since the dawn of humanity. The real difference lies in how we perceive them. In the meaning we assign. In the stories we tell ourselves. In the perspectives we see them from.

When we shift our perspective, everything changes. What once seemed like a roadblock can suddenly reveal itself as a doorway. What once felt like punishment may begin to look like preparation.

Contrary to what we've been taught, perspective isn't something you have to force through constant mental effort or repeat like a mantra. It's not about faking positivity or putting on a brave face when everything inside you feels like it's falling apart.

Real perspective, the kind that transforms you, isn't forced. It arises naturally when you get to the root of who you truly are. When you slow down enough to hear the quiet truth beneath the noise. When you stop trying to fix everything from the outside, and instead remember what you've always known but forgotten under the weight of cultural conditioning, trauma, and survival.

There lies true wisdom.

True wisdom is what we're missing, but what we so desperately need in today's world. We don't need more information; we're drowning in it. We don't need more intelligence; we're overflowing with it.

What we need is wisdom. The kind that has stood the test of time across generations, across continents, across civilizations. The kind of wisdom that doesn't bend with trends or break under pressure. The kind that lives in stillness, speaks through nature, and endures beyond logic.

We need that now more than ever.

Introduction

You didn't pick up this book by accident. There are no accidents in this universe. You're holding this book in your hands because you called it in. Somewhere deep inside, you've asked for change, for clarity, for meaning, for a way to come home to yourself. You want to be the one who sees life differently and who lives fully.

This book is not just the story of an ordinary woman who hit rock bottom, chose to rise, overcame what could have broken her, transformed her life, and now lives her dream. It is about the wisdom she uncovered along the way, the timeless truths that exist in nature, in light, in the atoms. It is about the courage to question everything society told her was true, and the discovery that the essence of life is not suffering, but joy, our original, natural state of being.

This book is a mirror, a compass, a guide. It will answer questions you didn't even know how to ask, ease the confusion that's been weighing on your heart, and help you remember what you came to this planet for. It will guide you to go deeper and explore the truth of health, of reality, of the self.

It will awaken your wisdom within.

These pages will take you not only to the grandest scale of existence, the universal and the planetary, but also to the smallest levels of life, the cellular and the atomic, to show you just how powerful you truly are, and remind you that you're life's greatest creation.

You'll discover that living a purposeful and joyful life isn't about pushing harder, enduring suffering with a smile, or managing your way through life. It's about aligning, remembering, and letting your dream life flow to you and through you, effortlessly, naturally, as it was always meant to.

ATOMIC GLOW

You'll realise that true healing and happiness don't come from bottles or devices, but from the power of you and Mother Nature. That clarity isn't something you chase, but something that emerges when you remove the noise. That the universe isn't out there somewhere, it's alive inside you, pulsing with intelligence, energy, and infinite potential.

It's your time to find truth at the deepest level of your being, to awaken the universe within, to radiate your atomic glow.

Let's begin.

CHAPTER 1

THE EARLY SCRIPT OF LIFE

"Life is a movie. Childhood writes the script, adulthood plays it."
Alison Davis

I can't count how many times my adult life has brought me back to the memory lanes of my early childhood. Sometimes it hits me like déjà vu, a moment, a scent, or a sound could take me back to something that feels eerily familiar, something that feels like it happened before.

Maybe it's part of getting older. Or maybe we're hardwired this way, designed to remember, to trace back, to make sense of our present through the lens of the past.

ATOMIC GLOW

A Childhood of Simple Beauty

Growing up in rural China, my early life now feels like a dream, a dream filled with simplicity, creativity, and resilience. We didn't have much, not by the world's standards, but we had everything we needed.

Each family in the village was allocated a small piece of land. From that land, we grew our food, like wheat, corn, and watermelons. We also grew cotton, which became the clothes we wore, the quilts that kept us warm in winter. We raised chickens for eggs and meat, and we dug a well for fresh water, cold, clear, and alive. Every element of life came from something close, something we could touch with our hands and shape with our effort.

Our days were simple yet purposeful. I'd wake up at dawn, eat breakfast, and walk to school, then return home to help with whatever needed doing. Sometimes it was foraging wild greens from the hedgerows, or carrying buckets of water from the well, which felt impossibly heavy in my small hands. I swept the yard, fed the chickens, and collected eggs.

There was no technology, no screens. Inside the house was, frankly, boring. And for my parents, at times, a battlefield. My curiosity knew no limits. I dismantled anything I could get my hands on, radios, clocks, and, much to my father's despair, his eyeglasses.

But outside was freedom. My brother and I, along with the neighbourhood children, lived our best lives in the dirt and wild. We played with sticks and pebbles, jumped off walls and rocks that led to scratches and bleeding, and climbed trees that often made my mother scream with horror. Even now, as an adult, I can hardly pass a big tree without the urge to climb it, like a call from an old friend.

The Early Script of Life

Life was quieter then, less noise, fewer distractions, and no rush to be elsewhere. And somehow, it was full. There was more presence, more participation in the moment, and more awe at the little things: rolling on the ground laughing after disappearing in the piles of autumn leaves, or running around trying to catch snowflakes in the mouth. We didn't realise it then, but we were living in a kind of abundance most of the world has forgotten.

Root of Creativity & Resourcefulness

My parents made nearly everything we needed. Meals were made from scratch. Clothes were sewn by hand or passed down. Furniture was basic, often built by my father. Even our house was built from clay and timber by his hands.

Electricity was a late addition to our village. We had no fridge, so my mother preserved food with age-old techniques handed down from her mother, and hers before that. She even made vinegar from scratch. And to this day, I can still feel the taste of the tangy, sour liquid straight from the vinegar mill when I come home from school. It often made my eyes twitch and my teeth grind, but it was one of my favourite things in the world.

A simple life without too much stuff taught me two instrumental values that underpin how I navigate life to this day: creativity and resourcefulness. They're reflected in my cooking; I always find substitutes for ingredients we don't have. They're reflected in my problem-solving skills; there is always a way! When it's a dead end, I make a way. They're reflected in my parenting, in my relationships, and everything. I learned to trust my hands, my instincts, and my heart. I can find ways to live a good life, no matter what.

Story of a Ugandan Teacher

What saddens me is that this simple beauty is largely lost in modern life. We are surrounded by so much, and yet taught to believe we don't have enough. Not enough tools. Not enough time. Not enough knowledge. Not enough things.

A teacher from Uganda once reached out to me via social media for a small donation to help his students. I felt the calling to chat with him over the phone. He explained that his students were so disadvantaged. They didn't have enough pencils and paper, there was no playground where they lived, and they couldn't even afford to buy shoes.

I couldn't help but ask him why he thought they were disadvantaged. He responded that they were below the standard, not living up to expectations. I explained to him that when I was a kid, I used branches to draw and write on dirt and sand. I explored the biggest and most fun playground of all, Mother Nature. And while my mother handmade shoes for us, most of the time we went barefoot because we didn't want to get them dirty. In fact, even now, I own expensive shoes, but most of the time, I don't wear any. There are extensive studies showing that grounding has tremendous health benefits. Why block those benefits with rubber-soled shoes?

After listening to me, he was silent. It was a new perspective he had never heard. Finally, he said he would do some research, and I never heard from him again.

I often wonder, if I hadn't grown up that way, would I live as simply and connected to nature as I do now? Probably not. That childhood shaped me, not just my values, but my instincts. They drive how I live my adult life. They shape my character, how I perceive the world, and how I think, feel, and behave.

The Early Script of Life

The Scarce Money

Not all the scripts I wrote in childhood ended up serving me later in life. Let me share a big one I'm still mastering, even today.

Being raised in a family that didn't have much, we were taught to appreciate what we had, and that was deeply valuable. We were grateful to have enough food on the table and warm clothes for the winter. But one thing we didn't talk about was money.

Not at home, and not at school either. It was as if money wasn't something appropriate to talk about, let alone something to want more of.

The message I observed and heard the most was: save what you can, but never have too much. People who had a lot of money were seen as greedy or morally questionable. On the other hand, living with less was a badge of honour, a sign of being a good, loyal citizen.

That belief got embedded so deeply in my subconscious that it ended up guiding many of my decisions as an adult.

Scarce Money Playing Out in Real Life

After China began opening up in the 1980s, new jobs became available. My family eventually moved to the city so we could access better education. After graduating from college, I landed a job at the only 5-star hotel in the area, the Crowne Plaza.

Earning my first paycheck, just a few hundred yuan, was a big deal. I felt both proud and nervous. That was a lot of money. But deep inside, I didn't feel right keeping it. After all, "good people" didn't hold onto money; they gave it away.

So I spent all of it on the family. I bought our favourite buns as treats. I purchased a fridge, a wardrobe, a dresser…and I felt amazing. I was praised for being a responsible daughter who contributed to the family. That's all I ever wanted to feel valued.

This pattern followed me when I moved to New Zealand a few years later. Without my family around, I became even more frugal, saving every dollar I earned and only spending on the bare essentials.

When I finally had enough saved to return home, I spent it all on gifts, something for everyone: aunties, uncles, cousins, grandparents. I wanted them to know I hadn't forgotten where I came from. I even gave my parents back double what they had paid for my tuition. It was my way of saying, "Look, I'm still a good, loyal daughter. Even with plenty, I haven't changed."

Moments of Change

For years, that mantra guided me. Give. Prove. Provide. Until one day, it backfired. A family member asked to borrow money that I didn't have. And for the first time, the illusion of financial security shattered, the endless cycle of earning, spending, and pretending slapped me awake. That moment changed how I see money. I began saving seriously for myself.

The belief in saving more and spending little stayed with me even as I built a family of my own in New Zealand. Money is hard to earn. You might run out of it one day, so spend it sparingly. This was, without a doubt, the most limiting belief I carried about money, and it took years to unravel.

The Early Script of Life

Thankfully, my husband grew up with a completely different relationship with money. Where I saw lack, he saw flow. Where I tightened my grip, he opened his hands. Through his eyes, I slowly began to see money not as something to chase or control, but as something that could move with ease. I started to welcome it with love instead of resistance.

Still, that old voice of scarcity would knock now and then. It reminded me just how deeply these beliefs can bury themselves, and how stubborn they are when you try to let them go.

The strange thing is, in other areas of my life, that scarcity script doesn't apply. Even in moments of health crises, I trust my body's ability to heal. I can feel the abundance of vitality, even when the evidence says otherwise.

I realised, we each carry our own mix of strengths and blind spots. Some people effortlessly grow wealth but quietly suffer in their health. Others radiate vitality yet feel constant scarcity around money. The real power lies in recognising both. Honour what comes naturally to you and, with compassion and persistence, keep showing up for the parts that don't.

No Room for Fear & Doubt

Growing up wild and free, often without adult supervision, taught me to rely on my instincts without the distraction of fear and doubt. How far could I walk along the tree branches before they snapped? Was it safe to jump off the wall at a certain height? No one was there to tell me what to do. I had to decide, trust myself, and carry on. That early independence became my blueprint. When I set my mind on something, nothing could stop me.

A few years into working in the hotel industry, life began to feel complicated. The politics, competition, and mind games didn't sit well with me. I longed for change. So when the opportunity to study overseas came up, I shouted yes at the top of my lungs. I didn't care where. I just knew it had to be outside of China.

The agent mentioned New Zealand. I said yes without hesitation, without even knowing where it was or what it was like. I picked Lincoln University in the South Island simply because the name of the president sounded cool.

Getting to New Zealand

The story of how I got to New Zealand is nothing short of a fairytale.

Three days before the university's arrival deadline, my student visa still hadn't been issued. Everyone told me I might have to miss a term, but I kept packing my luggage and saying goodbye to friends, as if it were a done deal that I was leaving. The next day, the embassy called. My visa was ready.

I was thrilled, only to be told that all flights to New Zealand the next day were sold out. I didn't think much and took an overnight train to Beijing to pick up my visa from the embassy. Holding my long-awaited visa like it was a golden ticket, I boarded the return train home straight away.

At around 11 p.m., as I stepped out of the taxi near our apartment, I saw my uncle and aunt waiting by a car.

"Jump in," they said. "We're driving to Beijing right now."

The Early Script of Life

My luggage was already packed in the back. My parents stood beside it.

My father handed me a stack of American dollars.

"This is all we have for you," my father said. "It's enough for two years' tuition."

Seeing the confusion on my face, they explained: a plane ticket had freed up at the last minute, departing from Beijing the next morning.

Besides getting super excited about jumping on an airplane, my heart swelled; this wasn't just money, it was hope, trust, and every ounce of belief they had in me. I blinked back tears, hugged them tight, and jumped in the car. And off we went. Back to Beijing. The ride was a blur. I was still floating in the surreal feeling of leaving everything I knew to step into a brand new world.

I had no idea how to get to the university once I landed. No idea where I would sleep the first night. I knew no one. I spoke Chinglish.

But there was no fear. No doubt. Only pure excitement, and the wide-open wonder of what life was about to unfold.

Of course, I met a girl on the plane headed to the same university. She had already booked accommodation and offered to share.

The lesson? Life is full of miracles. When we live with trust instead of fear, the path opens in ways we could never predict. Fear and doubt don't protect us. They shrink us and close any door before we even try the handle. But when we let go of needing to know how, and just follow the yes in our heart, life rushes in to meet us, with people, places, and moments more magical than we could plan.

CHAPTER 2

YOU WERE BORN LUCKY

"The best luck of all is the luck you make for yourself."
Douglas MacArthur

Back to the Beauty of Simplicity

Landing in a rural town, in a quiet country, in the dead of winter was a shock to the system for someone from a city of millions where the lights never go out and the streets never sleep. Suddenly, I was surrounded by more sheep and cows than people. And when the Sun disappeared at 5 p.m., there was nothing but darkness and silence.

For the first week, I paced the tiny bedroom I shared with my new friend, back and forth, trying to make sense of it all. But soon I realised, there was nothing to make sense of. Nothing was happening, literally. Just stillness. Quiet. In that silence, I slowly understood, it was exactly what I needed. With no distractions, I could pour myself into my studies. As the days settled into rhythm and routine, life began to open up. Classes filled my mind, friendships began to form, and I started to appreciate the beauty in simplicity. A new kind of life was taking shape, one that wasn't loud or busy, but full in a different way.

Living Cost Solution

My university degree was a four-year program, and my parents had generously covered the tuition for the first two years. I quickly realised I needed money for living expenses too. Seriously, how did I not think of that before? On my first visit to the local vegetable shop, my jaw dropped at the checkout. Everything was so expensive. It was hard to believe, especially when I converted it back to yuan! At this rate, my second-year tuition would vanish in weeks. I had to find a way to support myself, and fast.

The very next day, a senior student who lived next door mentioned he was heading to a neighbouring town to apply for a part-time job at KFC. He asked if I wanted to come along. Of course I did! We caught the bus together and walked into the store. He confidently asked to speak with the manager. A kind man in glasses came out to greet us.

"How can I help you?" he asked.

"We want to work here," my friend replied. "Are you hiring?"

"Sure," the manager said with a smile. "Let me ask you a few questions and show you around."

Just like that, within a month of landing in New Zealand, I had my first job. It paid minimum wage, but that was all I needed. It covered my living costs and let me safely tuck away my second-year tuition. A big win.

When I told my family about my new job, they were proud and relieved.

"You're so lucky," they said.

Was it just luck? Or was it something deeper, something meant to be?

I've never thought of myself as "lucky" in the traditional sense. I wasn't handed shortcuts. But I've come to realise I am lucky indeed, because I create my own luck. I show up. I say yes. I keep moving, even when things look uncertain or unfair.

Luck, I've learned, isn't something that falls from the sky. It's something we cultivate in the way we think, act, and respond to life. Everyone is equally lucky. It's not about chance. It's about perspective. When we see that things are happening for us, not to us, life begins to open in the most unexpected ways.

Finding University Tuition

Once I figured out how to cover my living costs, the next challenge loomed: tuition. As an international student, fees were steep, around NZD 13,000 a year back then. I was fortunate to have come to New Zealand 20 years ago. With today's soaring costs, I'm not sure I could have filled the gap.

But when you set your mind to something, the impossible begins to shift.

Working weekends and sometimes weeknights at KFC not only gave me paychecks but also upgraded my social and language skills. I improved my English, learned about the New Zealand culture, and formed friendships beyond the classroom.

Looking back, it was one of the most vibrant and joyful chapters of my student life. Believe it or not, my popularity among friends soared, not because of my grades, but because I always came home with leftover KFC after my night shifts. Who knew fried chicken could win hearts faster than anything else?

At the same time, I was determined not just to survive but to thrive academically. I squeezed studying into every spare moment, including bus rides and breaks between shifts. Textbooks in hand, I made buses and back rooms my library.

That discipline paid off. I earned straight As in all my subjects and was selected as an Accounting Subject Tutor in my second year, a role that not only validated my effort but offered much-needed financial support. Every week, I was getting closer to bridging the tuition gap for the next two years.

In the fourth year of my bachelor's degree, I decided to pursue further studies and enrolled in a postgraduate honours degree, costing another NZD 16,000, which I did not have. That didn't seem to intimidate me at all. During the summer holidays, I landed a full-time job at a local accounting firm. I saved every dollar, and with a little help from my family, I made it happen.

A year later, I graduated with First Class Honours and was recommended to pursue a PhD with a full scholarship. While tempting, my desire to build a career and create financial freedom spoke louder. When the Big Four accounting firms came to campus to recruit, I applied and landed a job. With gratitude, I turned down the PhD offer and stepped into the next chapter of my life as an auditor.

A dream started taking root in the foreign soil. I've learned that no matter how high the mountain is, you can climb it. You don't need to know how; you just need to take the first step.

The Auditor Mindset

Joining one of the Big Four accounting firms was a privilege. Working as an auditor? Even cooler. One day, I was a small-town girl, the next I was in the city wearing tailored suits and driving a real car for the first time in my life. And by "real," I mean a car that didn't cough every morning or look like it belonged in a scrapyard museum. My previous two cars were dirt cheap and utterly unreliable. That was the scarce money mindset in motion, loud and clear.

Landing that job felt like a major win. I loved seeing the reactions when I told people what I did. "An auditor? Wow, that sounds intense. Are people scared of you?" Not quite. If anything, they're more annoyed than afraid, possibly the most consistent feedback we get in this line of work.

But I was good at it. Really good. Maybe it was in my DNA. Or maybe it was my endless curiosity. I wanted to understand how things worked, to dig deeper, to connect the dots, which are exactly the qualities a high-performing auditor needs. That drive helped me climb steadily through the ranks and, after three years, I earned my Chartered Accountant designation. Those years shaped the foundation of my corporate career and gave me credibility in rooms I once felt invisible in.

Still, I often found myself wondering: why auditing? Sometimes it felt dull, so far removed from the practical, hands-on world I admired. I envied teachers who could apply their skills while raising their own children, or plumbers who could fix leaks at home in minutes. But an auditor? Not exactly the first person you'd call in a real-life emergency, unless your emergency involved a missing invoice or poor internal controls.

Yet, little did I know, the core principles I honed in that role would later become my lifelines.

Scepticism & Root Cause Analysis

One of the golden rules in auditing is: if it's not documented, it didn't happen. In other words, don't trust; they verify. That mindset shaped my natural scepticism and my thirst for truth. I learned not to accept things at face value. If something didn't sit right, I asked questions. I dug deeper.

And then there's root cause analysis, one of the most powerful tools I took from auditing. When something goes wrong, we don't just slap on a quick fix. We ask why, and then why again, until we reach the core of the issue. Only then do we address the real problem, so it doesn't come back.

That method of thinking, searching for truth and solving problems at the root, turned out to be essential not just in my work, but in my life. Especially when health challenges came knocking. I wasn't interested in band-aid solutions. I wanted to know why things were happening in my body, what the underlying causes were, and how to fix them for good.

So, while auditing might not help me unblock a sink or soothe a toddler, it has trained me to become a relentless seeker of truth. And that mindset? It's priceless. Without it, my life could have unfolded in very different directions, and this book might never have been written.

The Speed-Dating Coincidence

Ambition and drive kept me busy. Between work and friendships, my days were full, motivating, fun, and financially empowering. For the first time in my life, I felt truly in control. I could buy what I wanted, do things my way, and not worry about asking for permission or approval. I had built a secure, independent life.

Almost all my friends were in relationships, but I was perfectly content on my own. I wasn't looking. I didn't feel incomplete. Then, one day, a friend called and said she found two discounted speed-dating tickets online.

"You and our other single girlfriend have to go," she insisted.

Before I could even respond, she added, "Too late. She already bought them. You're going."

Okay, I thought. Why not? I had nothing to lose.

We both forgot all about it until a month later, when the tickets suddenly resurfaced in conversation. They'd technically expired, but we were lucky enough to get a booking confirmed. I remember laughing; clearly, this wasn't a sign from the universe, it must be a coincidence.

It was a rainy Tuesday evening after a long day at work. Honestly, all I wanted was to stay in with my pyjamas on. Then my phone rang.

"I'm in a taxi, will be at the event in 10 minutes. Will see you there!" my friend said, full of energy.

Reluctantly, I pulled on a pair of jeans and a T-shirt. I didn't need to impress anyone. But I didn't want to let her down, so I grabbed my bag and jumped into a taxi.

It was a dimly lit, trendy pub filled with mood lighting. As I walked through the room looking for my friend, I smelled the perfume from across the room, saw women in their best dresses, flawless hair and makeup, and men in sharp suits. I felt a wave of unease rising. What was I doing here?

Then I spotted my friend, glowing as always in a beautiful outfit. She took one look at me and laughed, "Jeans? No makeup at all?"

I grinned. "You're welcome. I wouldn't even be here if not for you."

We both burst out laughing and started wandering into the adventure that was about to unfold.

Meeting the Man of My Life

That night's speed dating event was specifically for university-educated professionals. Yes, we were on a mission to meet brainy gentlemen only.

As the evening unfolded, the men rotated through tables, each with a short window to make an impression. It was fast-paced, slightly awkward, and surprisingly fun.

At the end of the night, each person was asked to write yes or no beside the names of those they spoke to. I ticked yes beside the name

You Were Born Lucky

Miles, an interesting, kind young man who, like me, didn't seem to belong in that room. He felt real.

I still remember how he nervously sat across from me, trying hard to stay calm. At one point, he accidentally grabbed my glass of champagne.

I caught it and teased, "Hey, don't steal my bubbles."

He turned red and got even more nervous, which, of course, made me laugh. I was pleasantly surprised to find out later that he ticked yes beside my name too.

Weeks passed before he reached out. He invited me for a drink after work. When I arrived at the café, he was standing outside, waiting. He greeted me with a warm smile, opened the door for me, pulled out my chair at the table, and paid the bill before I even noticed.

A true gentleman, I thought.

We didn't talk for too long that night, but it was fun and deeply comfortable. I walked away knowing he was different. I walked away knowing...he was the one.

Two years later, he proposed on the Great Wall of China when we were travelling in Asia, including my hometown in China, and of course, I said yes! Soon after, we tied the knot on a beautiful sunny day in a breathtaking vineyard on Waiheke Island in New Zealand. A new chapter had begun, one rooted in love, laughter, and a chance encounter that was anything but random.

Setbacks & Struggles

By now, it might sound like everything unfolded easily for me, one joyful leap after another, from finding a job to meeting my husband. What I haven't shared are the struggles and setbacks quietly pacing through that same timeline. And, there were plenty.

There were long winter nights when I stood waiting for delayed buses in freezing temperatures, snow soaking through my shoes, wondering if I'd make it home before midnight. There were times I pushed through relentless illness because I couldn't afford to miss work or class. My first car was hit and written off by a stranger. I accidentally plowed my second car into someone's fence and had to use weeks of wages to pay for repairs. I didn't even have the luxury of slowing down to feel sorry for myself. I just had to keep going.

There's something profoundly powerful about being alone in a foreign territory. Without family nearby to lean on, you stop considering the possibility of failure. The idea of "I can't" disappears. There's no one else to pick up the slack, so you simply do it. You figure things out, no matter what. You don't stop to wonder if something might not work; it has to work. There's no room for hesitation, no time for a plan B, just the laser focus of moving forward.

Looking back, I realise that many of those struggles didn't even register as struggles at the time. They just felt...normal. Part of life. Another thing to deal with. Another lesson to absorb. Another layer of strength to build. In a strange, beautiful way, they were shaping the kind of resilience I never even knew I had.

CHAPTER 3

THE MOTHERHOOD FALL

"When a mother falls, she awakens."
Alison Davis

The Dream of a Perfect Mother

As a woman raised in a traditional Eastern culture, becoming the best wife and mother wasn't just a dream; it was a quiet expectation and destiny imprinted into my DNA. I carried it unconsciously for most of my life…until everything began to unravel.

We were blessed with two miraculous creations, and I feel the weight and wonder of that blessing every single day. But nothing could have prepared me for the intensity of becoming a mother, the collision of pain and pleasure, joy and exhaustion, love and confusion, all crashing in at once.

Holding my first son on my chest was surreal. My body was wrecked after a traumatic birth, and my heart still carried the unhealed wounds of a miscarriage before him. Yet there he was in my arms, perfect, warm, tiny but whole, new but somehow familiar. His little heartbeats fast and steady against my skin; his breath, soft and certain. He was sleepy, yet full of life. I had never felt so fragile, yet so in control. So broken, yet so complete.

It was the deepest fulfilment of a dream I had carried for as long as I could remember. Becoming a mother was everything I thought it would be, only messier, more beautiful, more humbling than I could have ever imagined. And I believed, with every fibre of my being, that I was finally stepping into the role I was born to play: the perfect mother.

Shattered Expectations

While I now deeply honour the magic of a nature-enriched, simple life, the younger me, raised in a family with limited means, longed for something different. I dreamed of being the kind of mother who could provide more: more choices, more comfort, more financial ease. I wanted to give my children everything I didn't have growing up. I believed that with enough effort, planning, and love, I could create a version of motherhood that was smoother, brighter, and more in my control.

But my son had other plans.

He didn't sleep when he was "supposed to." He didn't want to feed when I was ready with full, aching breasts. He cried for reasons I couldn't decode. I swaddled, bounced, rocked, and paced the floor for hours, but nothing worked the way I thought it should.

The harder I tried to gain control, the more powerless I felt. What once made me feel competent—perfectionism and determination—suddenly became my greatest enemy. With every failed attempt to "fix" things, I sank deeper into the belief that I was doing it all wrong. I felt like a failure, crashing hard from the illusion of a picture-perfect motherhood.

Emergence of Fear & Doubt

The perfect mother I had imagined was fading fast. The woman I once was—fearless, confident, in total control—felt like a distant memory. In her place stood someone I barely recognised: exhausted, overwhelmed, and second-guessing every move.

My son's blocked nose felt like a crisis. His cry sounded like an alarm bell I couldn't shut off. I worried endlessly about what I was doing, what I wasn't doing, what I should've done, and what I might never get right.

I feared that one wrong decision could ruin his future. That one missed cue might limit his potential. I began to ask the question I never thought would cross my mind: what if I'm not a good mother? Let alone the perfect one I had dreamed of becoming.

These thoughts didn't come from logic. They surfaced from somewhere deeper inside, unfamiliar and uninvited. I kept asking myself: where is this fear coming from? Who am I becoming?

Before motherhood, I moved through life with a clear, laser focus. My goals were linear. My path was a one-way lane, only moving forward, with discipline and drive.

But motherhood slowed me down and stretched my perspective far beyond achievement and control. It forced me to see what I'd always missed: the vulnerable, messy, uncomfortable parts of being human.

It wasn't until later that I truly understood that motherhood, in the same breath that brought fear and doubt, also opened a door to the opposite end of the spectrum. Through the chaos and collapse, it revealed an inner strength I never knew I had. A quiet resilience. A deeper kind of love. A beauty that lives not in performance or perfection, but in presence.

That's the truth I was beginning to uncover: motherhood isn't about control. It's about self-expansion. It doesn't just test who you are; it shows you who you truly can be.

Fear and doubt aren't the enemies I thought they were. They're the necessary doorways to growth, guiding us back to our most powerful and loving selves.

The Collapse

After too many sleepless nights, my energy was running on empty. I became irritable, short-tempered, and increasingly negative.

Trying to help, my husband arranged for a baby sleep trainer. I explained clearly that I didn't want my baby left to cry. In my heart, letting a tiny human cry himself to sleep felt like the ultimate betrayal. I believed that trust was built in those early moments, through presence, comfort, and love, not abandonment.

But the trainer followed her usual method. She let him cry. And cry. And cry. I stood there helpless, my heart breaking with every sound of crying. Until I couldn't stand at all. I collapsed, physically and emotionally.

That ended the "training." I held my son close to my chest. Guilt, shame, and powerlessness washed over me. The joy I had once imagined in motherhood was slipping through my fingers, replaced by tears, complete exhaustion, and pain.

Denial & Disbelief

The next day, I found out the sleep trainer had told my husband she believed I was suffering from postnatal depression and I should see a doctor.

"What? Me? She must have lost her mind!"

I was furious. How could I be depressed? I was healthy. Driven. High-achieving. Resilient. Happy. And I built a life from scratch in a foreign country. I firmly believed depression would never be part of my story.

One of the greatest gifts of my childhood was my health. There was no hospital nearby, and we didn't need one. Mother Nature was our healer. We ate what she provided, lived by her rhythms, and relied on traditional Chinese medicine when needed. I never had a serious illness and never took modern medicine.

One of my father's guiding principles had always stayed with me: "30% of all modern medicine is poison. Be wise about what you put in your body."

As I grew into adulthood, I became deeply invested in health and wellness. I believed that if I stayed healthy, I could stay in control. And if I were in control, nothing could go wrong. So, I ate well, exercised regularly, and did everything "right" as I'd been taught. I thought I was bulletproof.

The sleep lady's comment felt like a slap in the face. An insult. A threat to my identity. But deep down, I knew I was unravelling. The joy in me was vanishing. I was yelling more. Crying more. Blaming more.

I was desperately trying to hold on to the pieces of my broken identity until one day, I looked out at the lawn, and the grass that was once lush and vibrant now looked grey. The whole world around me, once full of colour, looked grey. I realised my vision was fading. For the first time, I panicked.

A Turning Point

I had no choice but to see the doctor.

She asked a few questions, then looked at me with quiet seriousness.

"You have severe postnatal depression and insomnia," she said. "You need treatment immediately, before it's too late."

Not grasping what she was saying, I asked what I should do. She wrote a prescription and told me to take antidepressants and sleeping pills immediately.

Thinking of my father's words that 30% of modern medicine is poison, my whole being was screaming no. I asked if there was anything else I could do. She shook her head.

"No, dear. You're falling apart. This will save your life."

I paused...then gathered enough courage to ask one more question: "What's causing the depression and insomnia?"

She shrugged, "Could be anything. Maybe stress? You're a new mum."

I asked, "So...will the pills take that stress away?"

She stared at me, almost bewildered. "No. But they'll make your life a whole lot easier."

That was the moment I knew. As slow and foggy as my mind felt, my heart was crystal clear: this is not the path I want to take. As an auditor, I was trained to find the root cause and fix it so it never returns. I didn't want to numb the pain. I wanted it gone.

"Thank you, but no." I said to the doctor, "I want to fix this for good. I'm so tired of feeling unhappy...and I'm going to find the happy me again."

What I didn't realise then was that this moment was the beginning of something extraordinary, the first step on the path that would transform my life. It was the doorway to my healing, the opening of a journey I never could have imagined.

The Awakening

I went home and headed straight to our tiny office, my heart set on one mission: to find the real cause of my breakdown and fix it at the root. As I opened the cabinet to grab a file where I could document my journey, a book tumbled out and landed at my feet. *The Happiness Advantage* by Shawn Achor.

The word happiness caught my eye immediately. That was exactly what I was searching for. The strange thing? We'd owned that book for years, and I had no memory of it being there.

There are no coincidences in the universe. It was meant to be. I sat down and began flipping through its pages. And before I knew it, I was completely immersed. Every word felt like it was written for me, speaking directly to the version of me that had been lost.

I was struck by the simple yet profound message: happiness is not the result of success, but rather the cause of it. Happiness is a choice, something we can actively create, even in the darkest of moments.

What hit me hardest were the stories of people living through war, trauma, and devastation, and still, somehow, cultivating happiness. If they could do it, why couldn't I? Suddenly, my situation didn't seem as impossible as I had told myself.

That book didn't just give me hope. It gave me solid evidence. The proof that happiness was possible for me again. That this overwhelmed, sad, disconnected version of me wasn't permanent. That there was a way back to joy.

A big shift stirred deep within me. I closed the book, stood up with tears in my eyes, and whispered to myself: "Bring it on!"

The Next Right Step

By then, I intuitively knew the root cause of my health crisis wasn't food or fitness. I had been eating well and exercising regularly. I had ticked all the right boxes for years, and still, I was falling apart.

The Motherhood Fall

But I needed to be sure. So I followed the nudge and enrolled in an 18-month health coaching course with a certified institution in Australia. I threw myself into it, driven by urgency and fierce determination. I completed the course and graduated in record time, just seven months.

And as I crossed that finish line, my intuition was confirmed: the root of my illness wasn't found in diet or exercise. That confirmation both relieved and ignited me. Relieved, because I wasn't doing everything wrong. Ignited, because now the real work could begin.

The auditor in me, the part trained to dig, to question, to follow the trail until the true cause is found, was on fire. What was really causing my health crisis? What was beneath it all?

Before my maternity leave ended, I made the most radical decision of my life. I resigned from my secure, successful corporate career to pursue what I now saw as the most important mission of all: to uncover the truth behind my health breakdown and heal from the root.

This was no longer just about me. It was about my family, the happy wife my husband used to know, and the loving mother my son deserved. It was about every new mother who had felt broken but didn't know why.

The journey toward answers, real answers to health and life, had begun.

CHAPTER 4

GAME OF LIFE IN THE MIND

"Play their game, you live their story. Play your own, you rewrite destiny."
Alison Davis

Life has a rhythm of ups and downs, highs and lows, blessings and heartbreaks, failures and successes. On the surface, it might seem like we're all playing the same game. But if you look closer, you'll note something fascinating; even when people face the exact same situation, their experience of it can be completely different.

Paradox of Life

You see it all the time.

Two colleagues are made redundant on the same day. One goes home, devastated, frozen in fear, convinced this marks the end of their stability. They spiral, blame the company, and lose confidence in themselves. The other goes home just as shocked, but sees it as a sign from the universe. Finally, a chance to start that business they've dreamed about. That very night, they pull out a notebook and begin mapping their next chapter.

It's seen in trivial moments in everyday life. Two people go for a walk on the beach and see rubbish scattered across the sand. One is irritated, pointing fingers at whoever ruined the view. The other quietly picks up the rubbish, tosses it in the bin, and continues walking, grateful to have made the place even a little better.

It happens in life-threatening situations. A wildfire sweeps through a small town and burns down homes. One family, holding each other, says through tears, "We've lost everything." Another family, also holding each other, says, "We're so lucky. We still have each other. Everything else can be rebuilt."

Same situation, two entirely different realities. Why?

It's not the situation that shapes your reality, but the story you tell yourself about it, the meaning you assign to it. It's the lens through which you see the world. They dictate every decision you make, every reaction you have, every word you say. In turn, these behaviours shape the reality and experience in your life.

GAME OF LIFE IN THE MIND

The difference is in the mind. Your life doesn't happen to you. It is constantly being created through the belief systems distilled in the mind through years of emotional patterns and past experiences.

Change Begins in the Mind

Trying to change a behaviour without changing your mind is like trying to grow a new tree from a rotten root. It might sprout for a while, but it won't last.

You may temporarily force a behaviour with rewards or threats. But once that external pressure disappears, your mind takes back control, and you'll snap right back to your old ways.

Take something as "simple" as changing how you eat. Let's say you want to stop snacking on sugary treats and start eating healthier. Logically, it's not hard. Just choose the salad instead of the cake. Right?

Yet almost everyone I know has struggled to make it stick. Because it's not really a battle of the hands reaching for food, it's a battle of the mind. Unless you shift what's driving the craving, the beliefs, the emotions, the subconscious associations, it's like trying to swim upstream with a lead weight tied to your ankle.

Many people attempt to overcome this through external motivators, either rewards or threats.

Imagine I told you, "If you stop eating cake for 10 days, I'll give you $500 cash." Suddenly, you can stop.

Or: "If you don't lose 10kg in the next few months, you won't fit into your wedding outfit."

That's exactly how I got my husband to lose weight before our wedding. I ordered his wedding suit two sizes smaller than what he wore. Let's just say…it worked like a charm. But did it last? Nope. A few months after the wedding, the suit fit a little too snug again.

Without changing the story the mind is operating from, old habits quietly return, as if they had never left. This is why real change is hard. Because it doesn't start with the doing. It starts with "seeing", the willingness to look deeper and update the operating system of your mind.

You're Responsible for Your Life

I'll never forget that day. I was walking barefoot along the beach, the salty wind in my hair, listening to Super Soul Sunday. Oprah was interviewing Bishop T.D. Jakes when he said something that struck me deeply: "Your life is a mirror of what's inside of you."

I had heard that phrase dozens of times before, in books, talks, and podcasts. But for the first time, I felt it. It landed in a way that bypassed my intellect and pierced straight through to my soul. I finally truly understood its meaning.

I'm the cause of my reality. Not my in-laws. Not my husband. Not the baby sleep trainer. Not life circumstances. It was me.

For so long, I've been blaming life for being unfair and that everything was working against me. I had been trying to change everyone else, blaming others for my frustration and pain. They need to be kinder. They need to communicate better. They need to understand what I'm going through.

But the truth is, I had created the life I was living. Not consciously, of course. But through my thoughts, beliefs, and emotional reactions, I had been shaping my reality from the inside out.

That was a defining moment, one of those rare turning points in life where everything shifts in an instant. For the first time, I understood what it truly meant to take radical responsibility. There was no one else to blame, no finger to point, no excuses left to cling to. It was me. All of it.

And strangely, that realisation wasn't heavy, it was liberating. Because in the same breath, I realised something even more powerful: if I was responsible, then I was also capable. I wasn't a victim of circumstance. I was a creator, not only of the life I am in, but also of the life I longed for.

That single shift from victimhood to ownership, from blame to creation, changed everything.

The Thoughts that Run Your Life

What you think shapes how you feel, and how you feel drives what you do. What you do creates the experiences you live, and those experiences become your reality. Your actions also influence your feelings, which in turn shape your thoughts, reinforcing the way you think and feel the next time around.

This is the Think-Feel-Do Cycle, a powerful feedback loop that quietly governs nearly every moment of your life.

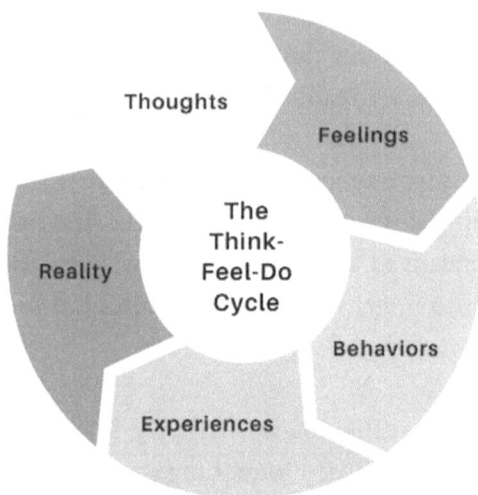

Graph 1: The Think-Feel-Do Cycle

Take someone who lashes out or constantly criticizes or blames others. It's not because they're a "bad" person. More often than not, it's because they feel misunderstood, disrespected, or threatened. So they act from a place of protection, driven by fear, frustration, or pain. Those feelings stem from deeper thoughts, like "I'm not being heard" or "People don't treat me fairly."

We saw this cycle played out in the story we explored earlier, where two people are made redundant on the same day. One person felt empowered and hopeful. They believed: "I'll figure this out," and took action to move forward.

The other felt bitter and powerless. They believed: "Life is unfair," and spiralled into blame. They told themselves a different story, and that story started with a thought.

To change the behaviour, you can either shift the thought that sparked it or the feeling that fuels it (which we'll explore in the next chapters).

Game of Life in the Mind

The Unconscious Thoughts

What most people don't know is that most of these thoughts don't even cross our awareness.

We have around 60,000 thoughts a day, and 95% of them are unconscious, habitual, and deeply conditioned. They arise automatically, without you choosing them. They run like background programs without you even realising they're there. These subconscious thoughts are what run your life, shaping how you see the world, how you feel, and how you respond. This is your subconscious mind, where your emotional patterns, beliefs, habits, and automatic responses are stored.

Only about 5% of your thoughts are conscious, the kind you're aware of, the ones you know you're thinking, like: "What should I make for dinner?" or "I need to reply to that email." These are deliberate and within your control. This is your conscious mind, where logic, decision-making, and willpower live.

The Battle with Your Thoughts

When there is a conflict between your conscious and subconscious mind, the subconscious always wins. You've probably experienced this battle between the conscious and subconscious mind before.

You know you shouldn't eat the sugary cake. You know you want to live a healthy lifestyle. But then, somehow, you're halfway through the cake before you even notice.

That's not you being weak, but your subconscious overriding your conscious willpower.

In your conscious mind (5%), you're making great decisions: "I want to be healthier. I don't need sugar. I'll feel better without it."

But in your subconscious mind (95%), there's a stronger, louder voice saying: "Sugar makes me feel safe. I need this to cope. This is how I comfort myself."

Unless you change what's stored in the subconscious mind, true, lasting change will always feel like an uphill battle.

When I Tried to "Think Positive"

I had gone to a mindset therapist once, hoping to shift my negative thinking patterns. I knew they were damaging my health, my marriage, and how I was showing up as a mother. I didn't want to be that way anymore; I just didn't know how to stop.

She introduced me to the "Thought Replacement" method. The idea was simple: take a negative thought and swap it for a positive one. For example, from "My in-law is being difficult" to "My in-law is helpful."

I wrote the new thoughts on sticky notes and plastered them on the fridge. Every time I caught myself thinking something negative, I was supposed to "replace" it with a positive one.

Sounds easy, right? Except it wasn't. It felt so false and forced. It didn't matter how many sticky notes I wrote or how many times I read them. My body didn't believe the words. My emotions didn't shift. Frankly, I wanted to punch the fridge. It was impossible. Because I was trying to use 5% of my willpower to fight against 95% of my subconscious programming. If you do the math, you'll understand, it's not a fair fight.

Game of Life in the Mind

How the Subconscious is Programmed

Your subconscious mind, the part of your mind that quietly runs the show, begins its programming long before you're even aware of it.

From the moment you're born to roughly the age of seven, your brain functions in a slow, hypnotic brainwave state called theta, the same brainwave state used in deep meditation and hypnosis. It's dreamy, absorbent, and open.

In these early years, your conscious mind isn't developed enough to judge, question, or think critically. There's no filter. You simply absorb everything as truth.

You don't decide what's right or wrong, or evaluate whether something is safe or fair. You don't have a mental firewall yet. You just download whatever is happening around you. That's how your early subconscious programming is formed, through absorbing the "rules" of the world:

- What's considered normal in a family or relationship
- What success, failure, or being "good enough" looks like
- What emotions are safe to express or to suppress
- What's expected of you to receive love or acceptance
- How love is given or withheld
- How safe (or unsafe) the world feels

You're not just hearing words. You're absorbing energy. The tension in a parent's voice, the look in someone's eyes, the silence after an argument, all of it becomes part of your program coding.

You internalise everything as if it's about you. If mum is stressed, you might subconsciously believe "I must be causing it." If dad never says "I'm proud of you," you may absorb "I'm not enough." Even if no one explicitly says these things, the subconscious mind picks up meaning through repetition, tone, body language, and emotional cues.

The Inheritance of Subconscious Thought

The programming doesn't start with you. Your parents and caregivers were also operating from their subconscious scripts, downloaded from their parents, and so on. So your subconscious is not just your own, it's a hand-me-down. You inherited beliefs about money, health, love, safety, worthiness, and success from those who came before you, often formed generations ago during times of survival, even poverty, and war.

That's why about 80–90% of those unconscious thoughts are negative and repetitive. The subconscious is designed to protect you, not to make you happy. Its job is survival, not growth, joy, or success.

If you were programmed in an environment where love was conditional, scarcity was normal, or emotions weren't safe, your subconscious is likely wired to anticipate pain and avoid risk, even when those things no longer exist.

This explains why so many of us in adulthood feel stuck or conflicted even when we consciously want something different. You might tell yourself, "I want to feel calm," but your subconscious has been programmed for stress. You might say, "I want to earn more," but deep down, you believe money is bad or unsafe.

In Chapter 1, I shared how I unconsciously inherited a "money is scarce" belief from my family. That program was quietly running in the background, influencing my decisions and relationships, until I became aware of it.

Game of Life in the Mind

Here's where everything changes. Just because your subconscious was programmed without your awareness doesn't mean it has to stay that way. You can rewrite the old script. You can consciously install new scripts, habits, and emotional responses that support the life you want to live. It starts with awareness.

CHAPTER 5

DEEP INSIDE OUR UNIVERSAL MIND

*"Your mind is an ocean. You either
master the tide, or be pulled under."
Alison Davis*

The subconscious mind is incredibly powerful and remarkably cunning. It drives almost every thought you think, every emotion you feel, and every decision you make…all while hiding beneath the surface.

Understanding the Subconscious Mind

If your conscious mind is the captain of the ship, your subconscious mind is the vast ocean underneath. You might believe you're steering the course, but what lies beneath, the unseen currents—your emotional memories, belief systems, and conditioned habits—are what determine whether you move smoothly or crash into a storm.

It doesn't just influence your life; it creates it.

As we explored in the last chapter, your subconscious mind is mostly programmed in your first seven years of life, when your brain is operating in a hypnotic theta state, highly suggestible and totally unfiltered.

It stores what you see, hear, and feel from your parents, caregivers, culture, and society as truths. It stores everything you've ever experienced, and every emotion, belief, and perception you've ever had. It's from this storehouse that your automatic thoughts arise, of which 80–90% are repetitive and negative.

It's the reason you react before you think. It's why certain things trigger you with no logical explanation. It's why you sometimes know better…but still can't do better.

Fast forward to adulthood, those unexamined childhood programs become the lens through which you view the world. You're not responding to the present moment, but simply replaying the programs you've stored over and over.

- You avoid setting boundaries because conflict feels unsafe
- You stay in draining jobs or relationships because change feels too risky
- You compare yourself online because you feel you're never enough
- You chase success to feel worthy of love
- You scroll late at night to escape the overwhelm

- You suppress emotions because being emotional is a sign of weakness
- You reach for sugar when stressed because it reminds your brain of comfort
- You seek validation to feel safe, even though it never lasts

You're not making these choices on purpose. You're simply operating from subconscious programming. The subconscious repeats what it knows, not because it's true, but because it's familiar.

Awareness: The Key to Change

To change the programs running your life, you must first become aware of them. Because the truth is simple: you can't change what you can't see.

The subconscious mind is like a train running on autopilot, repeating the same route over and over again. Unless you become aware of it, it just keeps going—same tracks, same destination, same experience.

The moment you shine a light on that hidden track, those invisible thoughts and old stories, you gain the power to slow it down and choose a new direction.

Pay attention to the quiet whispers in your mind, the automatic thoughts that pass through without resistance, the old, outdated programs installed long before you had a choice:

"I'll never get this right," when you're handed a new opportunity.

"They probably won't like me," before you even walk into the room.

"I don't have time," the moment something good tries to enter your life.

"I always mess things up," before you've even taken the first step.

As Viktor Frankl once said, "Between stimulus and response, there is a space. In that space is our power to choose our response. In our response lies our growth and our freedom."

That space is awareness.

It's the split-second moment where you pause and notice: "Wait…this isn't the truth. This is just conditioning."

That one pause, one breath of awareness, is the seed of transformation. In that pause, you reclaim your power. You shift from being reactive to being conscious. From living on autopilot to living by choice.

That is where change begins, not with force, but with awareness. Awareness is the first victory over the program. It doesn't end the pattern, but it opens the door to something new.

Living Here & Now

Awareness can only be practised in one place: the present moment. To access the present moment, the mind must slow down enough for you to see what's really going on inside.

But for most people today, the mind isn't just busy, it's racing. Planning, worrying, overthinking. Rehearsing what to say. Regretting what was said. Projecting what might go wrong. Replaying what already did. It's a constant mental time-travel, flipping back and forth between the past and the future, leaving the now completely unlived.

Deep Inside Our Universal Mind

Motherhood revealed this to me in the most confronting way. My son's simple cough was never just a cough in my racing mind. It instantly pulled up everything I'd ever read about coughs and childhood illness—the symptoms, the risks, the long-term consequences. I saw terrifying stories of complications, worst-case scenarios, and irreversible damage flash before my eyes. In a split second, I was convinced his future was at risk.

I was Googling late into the night, calling doctors, mixing remedies, desperate to fix something that didn't even need fixing. It was a normal cough. The kind every healthy, living human has from time to time. But I couldn't see that. I wasn't present with what was. I was reacting to a fear rooted in the past, and a worry projected into the future.

This constant mental time-travel keeps you locked out of awareness. The faster the mind races, the harder it becomes to catch your thoughts. It's like trying to read billboards while speeding down the highway; you miss almost all of them.

To become aware of the subconscious patterns driving your life, you have to slow down. You have to find stillness.

Right here, right now, is where you notice the thought, where you interrupt the pattern, and where you claim the power to choose something different. Because the present moment is the only place where awareness lives. The past is already gone. The future hasn't happened yet. Now is the only moment you can rewrite your story and create a new life. Now is where healing begins.

ATOMIC GLOW

Meditate & Slow Down

Meditation is one of the most powerful tools for slowing the mind and reconnecting with the present moment.

When most people think of meditation, they imagine sitting cross-legged, eyes closed, doing absolutely nothing for long stretches of time. For many, that feels daunting, boring, or even pointless. But that image is just one expression of meditation.

The word meditation actually means "to become familiar with." To become familiar with your thoughts. To observe your emotions without being ruled by them. To gently witness your behaviours, patterns, and inner world, not to judge or fix them, but simply to notice. Meditation is about coming home to yourself.

That means meditation can take many forms. Breathwork. Journaling. Silent reflection. Guided visualisations. Even walking in nature or tending to a garden can become a meditative practice if you are paying full attention. Because anytime you're fully aware of what you're doing, you're reconnecting with the present moment. And when your mind is anchored in the now, it naturally begins to slow down.

Frankly, meditation was not something I ever imagined myself doing. Years before becoming a mother, a friend took me to a meditation class. I could barely sit still, let alone take it seriously. I remember giggling, glancing around the room, and thinking, Why can't these people find something better to do?

But life has a way of teaching us. When I became a mother and my mind felt like a storm I couldn't quiet, I intuitively turned to Transcendental Meditation. I didn't fully understand it at the time, but something in me knew: I needed stillness.

It was in that stillness that I finally started to see the storm inside. The tension. The exhaustion. The thoughts I'd been running from. I began to hear myself again beneath the noise of the world, beneath the fears and expectations. I began to heal.

Turn Your Life into a Meditation

Over the years, my meditation practice has evolved through many forms. I started with Transcendental Meditation during early motherhood, then explored energy ball channelling, journaling, and guided meditations. And finally, I discovered my most profound and joyful practice: gardening.

Gardening not only keeps me rooted in the present moment, it connects me deeply with Mother Nature. It grounds me, heals me, and teaches me the timeless wisdom of life. It's become my sacred space. My therapy ground. A place where the mind stills, and the soul speaks.

But meditation doesn't have to look like gardening. Everyone resonates with something different. For some, it might be breathwork or yoga. For others, journaling, painting, dancing, walking in nature, or sitting in stillness.

The key is to find something that allows you to drop into presence, that quiet space between doing and being. Ask yourself:

- What brings me peace?
- What's sustainable and joyful for me?
- What fits my values and my rhythm of life?

Some seasons may call for longer practices, an hour or more a day, when you're deep in healing. Other times, five to 10 intentional minutes may be all you need.

The ultimate goal of meditation is not to escape life, but to become one with it. When you turn your life into a meditation, the lines between practice and living begin to blur. You don't have to find time or ways to be present, because presence becomes the way you live.

The Hidden Power of the Universal Mind

Your subconscious mind isn't just a personal storage unit of beliefs and patterns; it's also known as the universal mind, the quantum field, or, as Dr. Jill Bolte Taylor calls it, "Our Mind." It's like a computer with unlimited potential. Not a regular one, but a quantum computer, infinitely powerful, interconnected, and dynamic.

This universal mind is the field that we all share. It holds every potential, every idea, every solution, and every version of reality that has ever existed or will ever exist. It's a cosmic library of possibilities, a source of unlimited intelligence and creative energy, and your subconscious taps into this field like a radio receiver, tuning into frequencies that match the beliefs and programs you carry within.

Let's break that down.

The beliefs stored in your subconscious act like code, programs that filter and shape which possibilities collapse into reality. If your subconscious is programmed with "I am abundant," then that code filters reality to bring abundance into your experience. Opportunities, connections, ideas, and resources that align with abundance begin to appear.

But if your subconscious holds the belief "Life is unfair," then the quantum field will begin to match that belief too. It will collapse possibilities that reinforce injustice, struggle, and disappointment, because that's the command it was given.

It's like having a genie in your mind. Whatever you think, consciously or unconsciously, it replies, "Your wish is my command." Not just once, but continuously, because the subconscious doesn't judge. It doesn't distinguish good from bad, right from wrong. It simply delivers experiences that match the signal you're broadcasting.

You're already manifesting or creating all the time, not by effort or desire, but by default. If you don't take charge of the program, the program will run you.

Everything is Possible

Here is the most exciting part. Because the subconscious mind taps into the field of infinite possibility, anything is possible. Truly anything.

The airplane, the smartphone, the electric car, and the internet, all once "impossible" ideas, were made real when someone believed they were possible and aligned their thoughts and actions accordingly.

Even the solution to your deepest challenge, no matter how impossible it seems, is already out there in the field, waiting to be tuned into.

So what does this mean for you? It means that when you're stuck, there is a way through. When you're lost, there's a path forward. When you feel powerless, there's untapped power within you. The solution may not be visible yet. But it exists.

When you shift your internal code and begin commanding the subconscious with clarity and intention, your universal mind will respond, not because you forced it, but because you finally aligned with what was already available.

Your subconscious has no limit. It's a portal to limitless creation.

The Power of the Unknown

Most people fear the unknown. Uncertainty can often feel like a threat to safety, control, and comfort. The mind craves the familiar, even when the familiar isn't what we want. That's why so many stay stuck in unfulfilling jobs, unhappy relationships, or limiting habits, because at least they're known. Predictable and safe.

But the unknown is where dreams come alive. It's the field of infinite possibilities, the quantum field waiting to respond to your upgraded beliefs. It's the garden where every new reality begins to sprout.

Let's say you're craving a new job that lights you up, aligns with your purpose, and values your talents. Your current job drains you. It's unsatisfying, but it's known. You know the paycheck, the people, the routine. The job you truly desire lives in the unknown, an unopened door, an unrealised potential.

When you broadcast, "I am worthy of meaningful work," or "I trust life to support my highest path," your mind begins to scan the field of infinite potential for experiences that match that new code. Suddenly, doors open, and the right people appear at the right time. Your dream job that used to be in the unknown becomes the known.

This is creation. And you are the one who holds the power.

So the next time you feel fear in the face of the unknown, remember this: That same space you're resisting holds the very thing you've been yearning for. A new opportunity. A breakthrough. A miracle.

The unknown is not your enemy. It is the playground of your soul. It's where wishes are granted, where dreams are born, where the impossible becomes real. When you embrace the unknown with open arms, you don't just survive uncertainty; you command it as you wish.

When Change Fails: The Unsustainable Path

Now that you're aware your subconscious programs are shaping your life, the natural question becomes: what do I do about it?

Most mindset teachers will suggest one of two things:

- Replace the old, unhelpful programs with new ones that serve you better.
- Rehearse your desired beliefs repeatedly until they take root in your subconscious.

Some also recommend tools like meditation, visualisation, or hypnosis, methods that guide your brain into slower brainwave states (theta), which is how the subconscious was programmed in the first place.

This advice isn't wrong. As Dr. Joe Dispenza says, "Neurons that fire together, wire together."

The idea is simple: if you stop running the old program and start feeding a new one, the old neural pathways weaken, and the new ones strengthen and eventually take root.

I've tried them all. But the problem is, while helpful to a degree, they didn't always work and felt frustrating most times. Here is why.

Using 5% of your conscious willpower to fight 95% of deeply embedded subconscious programming is not a fair fight. It's exhausting. Remember the story I shared in the last chapter about trying to "think positively" about my in-laws? It felt like pushing a boulder uphill daily.

If changing one thought is that hard, imagine trying to manually rewrite the thousands, if not millions, of programs installed in the first seven years of your life. Many of which you can't even see. Some thoughts you think are the problem may only be symptoms. The real root might be buried deeper. For example, the thought, "My in-law is difficult," might actually be rooted in the belief, "I'm not safe when others disapprove of me."

So what do we do when rewiring thought by thought isn't sustainable?

Relationship with Fear

I found a clue much earlier in my life, before children, before burnout, before I ever thought about rewiring my mind. Remember the stories in Chapter 2? The "impossible" things I envisioned just…happened. Not because I reprogrammed each limiting belief, but because I created without fear and doubt. In truth, I didn't even have any belief systems about them. I just followed the vision and did it.

That's the key. Fear is the silent architect of almost every limiting belief you have.

Strip back the layers of worry, stress, insecurity, and at the core, you'll often find fear. Fear of failure. Fear of judgment. Fear of not being loved. Fear of the unknown. Even doubt is just fear wearing a different mask.

Once I became a mother, those fears surged back in tidal waves. I had to face them and learn to change my relationship with them.

Here's the radical truth:

The most sustainable, powerful way to reprogram your subconscious mind is not by editing it line by line like a giant spreadsheet. It's by shifting your entire relationship with fear.

When fear loosens its grip, the subconscious opens like a vast horizon, revealing infinite possibilities, untapped potential, and new ways of being. In the next chapter, you'll glimpse that power through the heart, another gateway capable of connecting you to the limitless field of creation itself.

By the end of this book, you'll understand a profound truth that changed how I live my life: All beliefs, even the "positive" ones, are limitations. They shape what you can see, but block what you can become. Because every belief defines and confines your reality. Your deepest dreams live in possibilities, the unknown that's beyond belief.

So for now, hold that thought. We're just getting started.

CHAPTER 6

THE POWER OF A BROKEN HEART

"The heart does not shatter when it breaks. It expands."
Alison Davis

Daily meditation became my lifeline. It didn't fix everything overnight, but it helped me slow the downward spiral I had been caught in for far too long. I remember those long, lonely days, sitting on the floor of our small room, door closed, trying to calm the racing thoughts. Many times I sat for hours. My legs would go numb, my back ached, and my mind fought me over and over. But a little voice inside whispered, "Keep going."

It wasn't easy, but it worked.

Bit by bit, light began to return to my life. I started to smile again. I noticed colour returning to the world around me. Laughter found its way back into the room. My son's giggles filled me with joy and hope. I was beginning to feel like myself again.

A Dream Child was Born

Since I wasn't working, and things felt more manageable at home, we thought, why not have another child? I finally felt like I had some rhythm with motherhood. Surely this time would be different. And it was.

Pregnancy with our second child was pure joy. I felt calm, grounded, and more connected to myself. I wasn't obsessing over every detail or outcome. I trusted my body. I trusted life. He was born with ease just as we'd hoped. He was a dream baby. Peaceful, content, feeding well, sleeping soundly. I thought to myself, this is how it's meant to be.

Mother Nightmare Returns

I would not have guessed in a million years that without warning, postnatal depression would return, fast and furious. No signs, no build-up. It just knocked me over. This time, it hit even harder.

With two young children to care for, the pressure was greater. The responsibility was heavier. The overwhelm was deeper. The angry, anxious version of me came alive again, only worse than before. My mind, which I had worked so hard to calm, began to unravel again. The spark I had worked so hard to reclaim vanished again.

I had filled my toolbox with all the right things: meditation, breathwork, healing practices, and mindset techniques. I had done the work. So why…why wasn't it working this time?

The Sleepless Body

To my horror, insomnia also returned.

If you've ever gone without sleep night after night, you'll know that nothing makes sense anymore. Everything feels hundreds of times heavier. Your energy runs on empty. Your patience wears thin. Your spirit feels like it's dying. Life becomes distorted and hope disappears.

My world began to collapse for the second time, and the familiarity was terrifying. I could feel myself sliding toward that same path I had once barely escaped, the deep darkness I vowed never to revisit. And yet, here it was again, pulling me under. For the first time, I realised how frightening it is when the body forgets something so natural, so fundamental, as sleep.

Why Sleep is the Foundation of Health

Sleep isn't just rest. It's repair. During deep sleep, your body does more healing work than you can imagine. Your brain clears out waste proteins through a system called the glymphatic system. Your muscles repair and grow. Your immune system strengthens. Your cells regenerate.

Sleep is also when your brain consolidates memories, fine-tunes your hormones, and restores emotional balance. During the night, your body regulates cortisol, the stress hormone that helps you stay alert and respond to challenges, and melatonin, the sleep hormone that signals your body it's time to rest and repair. When these two are in harmony, your body naturally shifts between an energetic state in the day and restoration at night. It's like a nightly detox and tune-up, resetting every system in your body: mental, physical, and emotional.

One of the key phases is REM sleep (rapid eye movement), which helps regulate mood and emotional resilience. Research shows that people who don't get enough REM sleep are more prone to anxiety and depression.

What Sleep Deprivation Does to You

Now imagine all of that…not happening.

When you're sleep-deprived, your stress hormone cortisol increases. Your immune function drops. Your blood sugar becomes harder to regulate. Your brain's prefrontal cortex, responsible for decision-making, reasoning, and emotional regulation, begins to shut down. Meanwhile, the amygdala, the brain's fear centre, becomes hyperactive. This means even small problems feel catastrophic.

Your body enters a state of survival, producing more inflammation, ageing faster, and increasing the risk of chronic illnesses. Scientists have linked chronic sleep deprivation to a wide range of conditions: anxiety, depression, weight gain, autoimmune issues, heart disease, and even Alzheimer's.

No wonder I felt like I was falling apart.

But what hurt the most wasn't the fatigue. It was the helplessness. The betrayal of my own body. The terrifying realisation that something so natural, so vital, was no longer accessible to me.

The Power of a Broken Heart

The Vital Question

What actually makes a person naturally fall asleep? This question became my obsession.

It was around this time that one of my most instrumental mentors entered my life, Dr. Jack Kruse, a pioneering neurosurgeon widely regarded as the Father of Light, Water, and Magnetism. He is also considered the first theoretical quantum biologist. If teachers like Dr. Joe Dispenza helped save me, then Dr. Kruse not only saved me, he transformed me.

From the next chapter onward, you'll see how his groundbreaking work in quantum physics and quantum biology brought me to the very core of what it means to live the best life possible, aligned with nature, light, and the fundamental laws of physics. While his teachings are rooted in deep science, I will explain them in a way that is accessible and practical, so you can understand and apply them without needing a scientific background.

The Body's Hidden Clock

Dr. Kruse introduced me to the science of circadian biology, the study of our body's internal timing system. Every process inside us, from hormone release to digestion to repair, runs on a precise schedule. This timing isn't random; it is set by signals from our environment. Our cells read light, darkness, temperature, and even the magnetic fields of the Earth as cues to know what time it is and what they should be doing. In other words, your biology runs not just on food or water, but on timing.

The first time I heard Dr. Kruse speak on a podcast, I was blown away by the heavy science, literally. I didn't understand most of it. But at the very end, the interviewer asked: "If there's one thing people should do for their health, what would it be?"

He replied, without hesitation: "Never miss another sunrise for the rest of your life."

That hit me like lightning. Maybe it was the nature part of me waking up. Or maybe it was the simplicity of it, something free, something available every single day, something I hadn't done in years.

That very morning, I opened my window and let the light in. Curiosity led the way, and what I began to uncover about sunlight and the laws of nature reshaped everything I thought I knew about health.

The Real Sleep Switch

Turns out, melatonin, the hormone that helps you sleep, isn't just about nighttime. It's actually set into motion at sunrise.

Morning sunlight, especially the UVA light, stimulates the production of serotonin in your body, primarily stored in your gut. A few hours after sunset, sensing darkness, serotonin gets converted into melatonin. That melatonin release is what makes you feel relaxed, sleepy, and ready for restoration.

So sleep doesn't start at night. It starts in the morning sun.

That realisation felt like someone lowering a branch to me while I was dangling off the edge of a cliff. I grabbed it.

The very next day, I went outside to watch the sunrise. Then the next. And the next. I have rarely missed a sunrise ever since. Soon, I was walking barefoot on the beach for hours every morning, letting the sunlight recalibrate my system.

Combined with breathwork and minimising artificial light at night, this simple practice became my lifeline. Slowly, my sleep returned naturally. That was one of the biggest wins of my life yet. Today, I sleep deeply and soundly, like a baby.

Soon, you'll discover that this is just the tiniest glimpse of how deeply sunlight influences your health and your entire life.

A Powerless Heart

Even though life was starting to look up—I began to have reasonable sleep at night, I was walking at sunrise, and I had moments of calm again—the wreckage left by two waves of depression had taken a quiet but devastating toll on the relationships closest to me. Most notably, my marriage.

The emotional withdrawal. The explosive outbursts. The miscommunications and misunderstandings. The blame. The anger. The silent cold war. On top of it all, I was spending hours each day meditating, walking along the beach, desperately trying to "fix" myself. Without realising it, I had placed an immense burden on my husband to hold everything else together. Slowly, my relationship with the person I loved most began to unravel.

One day, after returning home from my usual morning walk, I noticed something different. A heaviness hung in the air. I looked around and realised my husband's belongings were gone. Later that day, he called. His voice was distant and flat.

"I've moved out," he said. "I just couldn't do it anymore."

I froze in despair.

A broken marriage is one of the most heart-wrenching experiences a person can endure. But for me, it was more than heartbreak; it was the shattering of my identity, my destiny.

Coming from a culture where being a devoted wife and mother wasn't just expected, but essential to my worth, it felt like a complete collapse of who I was supposed to be. I wasn't just losing a husband, I was losing half of the mother with co-parenting. I was losing the identity of a "wife".

It shattered me in ways I didn't think I could survive. In devastating and unbearable pain, I fought, pleaded and hoped. But nothing worked.

I had no more strategies. No more tools. No more energy left to "fix" anything. I was out of options. For the first time in my life, I felt completely powerless.

The Miracle

There was nothing outside of me left to hold onto. Everything I thought I could rely on had slipped away, and the only place left to turn was within. I spent hours in meditation and endless walks along the beach, trying to find relief, anything that could temporarily numb the pain. But the truth was, I wasn't coping.

In my desperation, I reached out to a mutual friend I trusted, and I'm so grateful I did. Simply being accompanied and heard in challenging times is deeply healing. It wasn't something I would normally do.

The Power of a Broken Heart

Perhaps it was that deeper knowing, the kind you begin to access when you finally turn inward, that guided me to reach out.

To my horror, my sleep vanished again, and night after night I was gripped by panic attacks. My heart would pound so fiercely I thought it might leap out of my chest. Later, I would learn there is a name for this, broken heart syndrome.

One night, when the pain felt unbearable, I placed my hand gently over my chest and whispered, "I'm sorry, heart. I haven't treated you well. From now on, I promise to take care of you. I won't let you be hurt again." To my astonishment, something profound shifted immediately. The pain softened. My racing heart began to slow.

I sat in awe, speechless at the power, the sheer magic of a simple touch infused with pure love. In that moment, something stirred awake deep within me, a part of myself I never knew existed.

My friend was trying her best to help reunite us. One day, after a long conversation with my husband, she called.

"I'm sorry," she said. "I don't think there's a way around it. He's made up his mind. I think it's time you let go."

She told me what they had discussed. All the heavy memories.

Something unexpected happened at that very moment. Instead of reacting with anger or self-pity, I felt an overwhelming love for him. I saw how painful everything must have been for him to endure. All I felt for him was love, the kind I never imagined possible, especially after he left when I needed him the most. I just wanted him to be happy, to feel free.

I whispered to my friend, "I love him. I'll let him go."

That night, I lay in bed with love swelling in my heart. I felt grateful for the years we shared, for the memories, and for the two incredible sons we created together. I said goodbye in silence. With a prayer in my heart for his happiness, I drifted to sleep.

The next day, we were supposed to meet and begin the process of dividing our assets. I felt calm. I was ready.

He came. But to my greatest surprise, he wasn't there to separate. He spoke instead about the possibility of coming back together.

To this day, I don't know what changed. He couldn't explain it either. But I know, when a broken heart opens and radiates love, miracles appear.

The Antidote of Fear

That was the first time I truly understood the power of polarity, how adversity can transform into opportunity when we respond in a way that defies logic, bypasses belief, and transcends fear, through a simple willingness to act from the heart.

Letting him go didn't just free him; it freed me. From that space of surrender, new doors began to open.

Do the opposite. Feel the opposite. As St. Francis of Assisi once said:

"Where there is hatred, let me sow love.
Where there is injury, pardon.
Where there is doubt, faith.
Where there is despair, hope.

The Power of a Broken Heart

Where there is darkness, light.
And where there is sadness, joy."

The power of polarity never ceases to amaze me. Every time I choose love or joy in the face of fear and adversity, it transforms into something truly beautiful in ways I could never have imagined.

I realised, we are all born alchemists, capable of turning every setback into strength, every low into a high, every heartbreak into healing, simply by choosing love. Because love is our natural state, and resistance to it is what brings pain.

The heart always wants to open. We can either choose to let it, or life will do it for us. Heartbreak serves one purpose: teaching you to open it through experience, until we've learned to do it willingly.

Going Deeper

Understanding this truth didn't make it easy to live.

Meditation gave me fleeting moments of connection, glimpses of wholeness, of pure love. It helped me see the power of living from the heart. But opening the heart, when everything inside screamed to close it for protection, often felt like tearing myself apart. Calming the mind sometimes felt like a battle I couldn't win, requiring more effort than I had to give.

One day, after a long meditation, it hit me: none of this made sense anymore.

I came to Earth to experience life, and yet here I was, alone, in a dark room, trying to escape back to where I came from, wholeness, pure love.

ATOMIC GLOW

I remembered Eckhart Tolle's words: "Stress is being here, but wanting to be somewhere else."

I suddenly realised that even meditation, as still and sacred as it seems, can become another form of resistance, stress, or another way of trying to escape life itself.

This couldn't be the root cause I was searching for. To find true freedom, I had to keep searching, go deeper, into the source.

CHAPTER 7

THE ACTIVATOR OF THE ATOMIC UNIVERSE

"Let there be light."
Genesis 1:3

After my heart cracked open and the healing began, curiosity, like childlike wonder, took me to a whole new level. I found myself asking deeper questions than ever before.

How does the mind truly work? What makes a heart open or shut? What are they made of, not metaphorically, but literally? And what do they really need to thrive?

I began to see the mind and heart not just as logical and emotional centres, but as living instruments, finely tuned, powerful, and intelligent in ways I never understood. But like any instrument, I knew they couldn't function well without the essential needs.

These questions began looping through my awareness like a song I couldn't turn off. They pulled me further inward, past emotion, past behaviour, and even past the subconscious mind…all the way down.

Down to atoms.

What began as a quest for healing became a journey into the very fabric of reality, the raw materials not just of the brain and heart, but of life itself. The root cause of our suffering, the foundation of the biology of being human.

The Fundamental Forces of the Universe

Let's start with the vastness of the universe, where stars are born and galaxies are formed, before zooming all the way down to the world of atoms.

According to the latest discoveries in science, everything in the known universe, from galaxies and stars to planets and all forms of matter, is governed by four fundamental forces. These aren't just abstract concepts in physics; they are the invisible scaffolding of reality, holding the universe and everything in it together. They shape the structure of the cosmos, influence the dance of atoms, and ultimately govern life itself. We will explore their equally fundamental role in human biology in later chapters.

1. **The Strong Force** – This is the most powerful of the four fundamental forces. It's what holds the nucleus of every atom together. Without it, atoms, the building blocks of all matter, would fall apart. No atoms means no matter. No stars, no galaxies, no Earth. No you, no me. No physical universe. It's the glue of physical existence.

The Activator of the Atomic Universe

Application in biology: it ensures the stability of atoms that ultimately absorb and emit photons in biological systems.

2. **The Weak Force** – Often overlooked, this subtle yet essential force governs radioactive decay and subatomic transformation. It's the quiet alchemist of the universe, shaping change from within. Without the weak force, stars wouldn't shine, elements couldn't form, and the cosmos as we know it would not exist.

   Application in biology: The weak force supports your mitochondria, those tiny engines you will come to deeply appreciate, by acting like a subtle "switch." It allows protons and electrons to move and exchange places in just the right way. This delicate dance, known as proton tunnelling and redox reactions, is what makes energy conversion possible inside your cells.

3. **The Electromagnetic Force (EMF)** – This is the most familiar and influential force in the universe and your everyday life. At the cosmic scale, it governs the attraction and repulsion between charged particles, allowing atoms to bond and form everything from stars and lightning to planets and plasma. It's the reason light travels through space, why Earth has a magnetic field, and how energy moves across the fabric of the universe.

   Application in biology: EMF is what allows atoms to assemble into you. It directly ties into how cells use light for energy and how they keep precise circadian timing. It powers your heartbeat, fires your neurons, enables your vision, and fuels every thought in your brain. Every sensation, emotion, and movement is driven by electromagnetic interactions.

4. **Gravity** – The most well-known and most gentle of the four. It's weak at the atomic level, but across space and time, it shapes the orbits of planets, the birth of stars, and the structure of galaxies.

    Application in biology: It's what keeps your feet on the ground and gives your body its orientation. It also works together with light-dark cycles to influence circadian timing and biological rhythms.

Atoms: The Building Blocks of the Universe

Now let's zoom all the way down, from the vastness of galaxies to the tiniest building blocks of reality: atoms.

Everything you see, touch, and interact with in everyday life is made of atoms. The trees outside your window, the birds in the sky, your phone, the chair beneath you, your loved ones, and you. It doesn't stop with the visible world. The air you breathe, the stars light-years away, even the empty space between thoughts, are all made of atoms.

They are the foundational bricks of all matter, galaxies, planets, oceans, forests, and human life. And they're anything but static. Atoms are intelligent, dynamic systems, tiny bundles of energy and information, constantly in motion, constantly responding to their surroundings.

Atoms don't just exist; they interact. They absorb, emit, and transform energy, especially light, continually adapting to the world around them. In other words, the universe is alive with atomic intelligence.

And so are you.

The Activator of the Atomic Universe

Here's what a basic atom looks like:

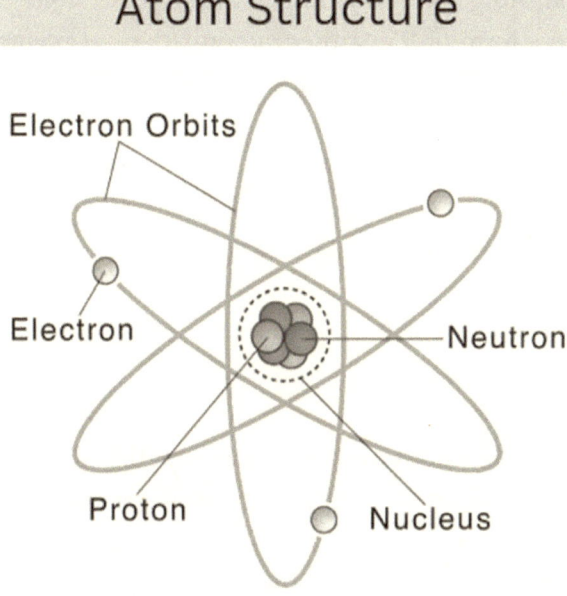

Graph 2: Structure of an Atom

At the centre of every atom lies a tiny, dense nucleus, made up of protons (positively charged) and neutrons (neutral). These particles are held tightly together by the strong nuclear force, the most powerful of the four fundamental forces in the universe.

Orbiting this dense nucleus is a cloud of electrons, each carrying a negative charge, moving within an invisible electron cloud, and held in place by the electromagnetic force, the most relatable force to our daily lives. This force enables atoms to interact, bond, and form everything from molecules to human thoughts.

ATOMIC GLOW

Currently, there are 118 known elements on the periodic table, each made up of a unique type of atom.

Isn't it astonishing that everything in the universe, from galaxies and stars to every form of life, arose from just these elements, endlessly moving, reshaping, and reorganising through laws of physics and chemistry?

The Astonishing Space Within

Here's the part that still blows my mind: the atom is mostly empty space.

If the nucleus of an atom were the size of a marble, the electron cloud surrounding it would span an entire football stadium. That means 99.9999999% of every atom, and therefore everything you see, including yourself, is made up of invisible space. But it's not empty in the meaningless sense. It's filled with fields of energy, force, and information.

This is one of the reasons physicists say most of the universe is invisible. The light that reaches our eyes only reveals a sliver of what actually exists. The majority of the cosmos, just like the atom, is energy we cannot see.

But here's what's truly fascinating: it's not the dense nucleus that most directly determines how an atom behaves; it's the orbiting electrons and their cloud. This seemingly insubstantial, invisible outer layer defines the atom's size, shape, reactivity, and how it bonds or interacts with other atoms. Changes in the electron cloud can even influence processes within the nucleus itself, while activity in the nucleus can subtly affect the electrons in return. Yet in the dance between the two, it's the outside that leads; the electron cloud sets the stage, drives the action, and shapes the story.

In essence, it's the invisible, dynamic forces on the outside that most powerfully shape what happens at the core, and the core, in turn, responds.

Atoms & Human Biology

In human biology, we rely on just about 25 of the 118 known types of atoms on the periodic table. These atoms bond to form molecules. Molecules organise into cells. Cells assemble into tissues. Tissues become organs. And together, organs create you, an intricate, intelligent, living system.

Yes, that's right. You are not just flesh and bone; you are trillions and trillions of atoms woven into patterns so sophisticated and coherent that consciousness, emotion, memory, and love can exist. You are a living symphony of vibrating atoms, orchestrated by the very same forces that shaped galaxies, stars, oceans, and the Sun.

You are not just in the universe. You are the universe, expressed in human form.

And here's what might be the most humbling truth of all: about 99.9999999% of what makes you is invisible.

It's not your muscles, your skin, or even your genes that define you. It's the invisible architecture of energy and information, the electromagnetic interactions between atoms, the subtle quantum fields, and the coherence between light and biology, that gives rise to your visible life.

So the next time you look in the mirror, remember: you are stardust, animated by light. A cosmic being in a biological body. And you are more powerful than you've ever been told.

Fun fact: Because of the electromagnetic forces between atoms, specifically the repulsion between the negatively charged electron clouds, technically, nothing ever truly touches. What we perceive as contact is actually the interaction of these invisible electromagnetic fields. There is always a tiny space at the atomic level. Solidity is an illusion.

Everything Is Energy

The 99.9999999% invisible space filled with forces and information at the outer layer of an atom is what we call energy.

Energy is the capacity to do work, to move, change, interact, and transform. It's the universal currency that powers all action and creation, from the dance of galaxies to the beat of your heart.

On the cosmic scale, energy governs how stars shine, how planets move, and how light travels through space. At the biological level, it is the spark of life itself. It is what keeps you alive. It powers every heartbeat. It fuels every thought and brainwave. It enables DNA to replicate, wounds to heal, and cells to communicate. It is the reason you breathe, blink, move, and feel.

But there is more. Let's zoom in even further, into the remaining 0.00000001%, the seemingly "solid" nucleus made of protons and neutrons. You might imagine them as tiny balls of matter, but quantum physics reveals something very different.

The Activator of the Atomic Universe

These particles are not solid. They're actually bundles of vibrating energy, behaving as both particles and waves depending on how they're observed. It's vibrating movement, energy in form. In other words, matter is just energy condensed to a slow vibration.

Albert Einstein captured this truth with his iconic formula: $E = mc^2$, where:

- E is energy
- m is mass
- c^2 is the speed of light, squared

This elegant equation changed everything. It tells us that mass and energy are not separate substances; they are two forms of the same thing. The "solid" matter you see and touch? It's just energy held in form, condensed through the speed of light.

So yes, everything is energy.

Just like the atoms, the universe, and you are fields of intelligent energy, with 0.00000001% wrapped in condensed form and forever connected to the cosmos.

Different Forms of Energy

Energy shows up in many forms, all playing vital roles in how the universe and your body function. The most relevant forms in both physics and biology include:

Light Energy

This is the most foundational form of energy, powering life at every level of biology and ecology. It includes:

- Photonic Energy – This refers to the full spectrum of sunlight that reaches the Earth's surface, including both visible light and invisible wavelengths such as ultraviolet (UV) and infrared (IR). IR, also known as thermal energy, is a form of radiant heat that your skin can feel but your eyes can't see.

As you'll discover in later chapters, human biology has evolved to not only respond to these light frequencies but also to generate and emit certain bio-photonic signals internally, some of which mirror or extend beyond these frequencies.

- Electromagnetic Fields (EMFs) – These are fields generated by electrically charged particles. They exist in natural, biological, and artificial forms:

    → Natural EMFs – Originate from Earth's magnetic field and from cosmic sources such as cosmic rays (high-energy particles from space). These fields help regulate circadian biology and orientation in animals, including humans.
    → Solar EMFs Beyond Photonic Light – The Sun emits the full electromagnetic spectrum, including high-frequency radiation (e.g., X-rays, gamma rays) that is naturally filtered out by the Earth's magnetosphere and atmosphere. These higher-energy frequencies do not reach the Earth's surface in meaningful amounts, but they're part of the cosmic environment under which life evolved.
    → Artificial EMFs (Non-Native EMFs) – Over the past few decades, humans have developed technologies that emit isolated, pulsed, and persistent EMFs, such as those from home lighting, screens, WiFi, Bluetooth, cell towers, and other wireless technologies. These

artificial EMFs mimic certain bands of the solar spectrum, which are non-native to biological systems, and can disrupt the body's natural electromagnetic environment.
- → Biological EMFs – These are the electromagnetic fields generated by your own body, also called biofield, especially the heart and brain, which produce strong, measurable fields that influence both internal physiology and your interaction with the external world.

Chemical Energy

Stored in the molecular bonds of the food you eat. Your body converts this energy into ATP (adenosine triphosphate), your cellular fuel, to power everything from movement to brain function.

Gravitational & Kinetic Energy

These are movement-based forces:

- Gravitational energy grounds you to Earth, helping regulate your nervous system and balance.
- Kinetic energy is created through physical movement such as walking, dancing, exercising, activating circulation, metabolism, and vitality.

Why Light Is the Most Fundamental Energy

Among all forms of energy, light, also known as electromagnetic radiation, is the most fundamental and far-reaching.

Light is the universe's fastest, most intelligent messenger. It travels with no mass, at the speed of 299,792,458 meters per second, and carries energy and information across infinite distances. Photons, the particles of light, can transmit signals from stars billions of light-years away to your eyes in an instant.

At the universal level, light organises matter. At the biological level, light organises life.

Biology runs on light. Plants absorb sunlight to make food through photosynthesis. Your cells absorb light to make energy through cellular respiration and mitochondrial function. Light regulates your biological clock. It tells your body when to wake and sleep, when to repair and grow, when to produce hormones, and even when to feel joy.

Light sets the rhythm of life. It's not just a source of energy. It's a source of order, clarity, and coherence in both the universe and within you.

Light: The Activator of Ecology & Biology

"Let there be light."

This ancient passage, found in the opening line of the Bible, holds more truth than we ever imagined. Long before science caught up, ancient wisdom already understood: light is the origin of life.

Modern physics reveals why.

Light gives rise to the electromagnetic force, the most influential and life-relevant force in the universe. Light carries energy in the form of photons, and when photons interact with charged particles, like electrons, they create electromagnetic movement. As light travels,

The Activator of the Atomic Universe

its electric and magnetic fields oscillate at 90 degrees to each other, generating a dynamic, interactive electromagnetic force.

That's why electrons orbiting the nucleus are key to atomic life. Electrons don't interact with neutrons or protons directly. They only interact with light.

Without light, electrons would remain inert, uncharged, unmoving, and there would be no electromagnetic force to spark reactions, build molecules, or generate life.

No light, no spark. No spark, no life.

Light is the first domino in the grand cascade of ecology and biology.

But so much of modern health focuses on the dominoes far downstream, biochemistry, supplements, symptoms, while ignoring the initial activator that sets everything in motion. Trying to change biology without light is like trying to play music on an electric piano with no electricity.

The very first and most important thing I learned from Dr. Jack Kruse was this: "Life is about light." I remember hearing those words and thinking, How could something as simple as light be the answer? It wasn't until I dove into quantum physics and studied the structure of atoms that it all clicked.

Of course, it's light. It had to be.

For centuries, Newtonian physics taught us to view the body as a machine—mechanical, separate, and purely physical. But that model ignored the invisible energy that animates us.

ATOMIC GLOW

Quantum physics changed that. It shows that everything is energy and matter, frequency and form, consciousness and code. It is the science that finally explains how life truly works, from the atomic to the cosmic, from biology to spirit.

And I believe this should be the science we learn from a young age. Not just to understand the universe. But to truly understand ourselves.

CHAPTER 8

THE EVOLUTION OF LIFE ON EARTH

"Nothing in biology makes sense except in the light of evolution."
Theodosius Dobzhansky

After journeying deep into the quantum world, I began to see the universe not as a straight line through time and space, but as a spiral of infinite layers, interconnected, intelligent, and alive.

In the grandest view, only a tiny fraction of the cosmos is made of matter, yet within that fraction lie billions of galaxies. Within each galaxy are billions of solar systems. Within each solar system are planets, and within each planet, including ours, are elemental building blocks that make life possible.

And within each of those elements—even within the atoms—lie still more layers: particles, forces, fields, and consciousness, interwoven like a cosmic tapestry.

Earth: A Layer of The Cosmic Tapestry

Earth is just one layer in the vast spiral structure of the universe. But to us, she is everything—our home, our cradle, our great Mother.

Just like every other layer in this infinite cosmic spiral, Earth herself contains countless layers of life within her. Humans are one, so are the oceans, the mountains, the forests, the birds, the whales, the insects, the fungi, and the microbes. Each is a unique expression of life, a living layer of Earth's great body, playing its part in the harmony of the whole.

And just like Earth, you, too, are layered. You are not just a person, you are a vast, intelligent community of trillions of living cells, working together to form the miracle that is you. We'll explore this incredible inner universe in the chapters to come.

For now, let's zoom in on Earth herself, how she came to be, how she evolved into a living planet, and how she continues to sustain life through the most elegant and intelligent systems ever created.

Because understanding how Earth works is not just about evolutionary history. It's about remembering who we are, where we come from, and how we're meant to live in harmony with the larger whole.

The Evolution of Life on Earth

Formation of the Earth

Roughly 13.8 billion years ago, the universe burst into being through what we now call the Big Bang. In the first moments, light was the very first form of energy to emerge, flooding the cosmos. It wasn't until about 4.6 billion years ago that our Sun ignited, formed from the collapse of a swirling cloud of gas and dust in the Milky Way galaxy. As gravity pulled these elements inward, pressure and heat built until, boom, nuclear fusion lit the Sun like a cosmic firestorm.

From that moment on, sunlight became the guiding force of our solar system. It warmed the planets, shaped orbits, and illuminated the space where Earth would soon take form.

Earth was formed about 4.5 billion years ago. A young, fiery planet, Earth began as a chaotic ball of molten rock, constantly bombarded by asteroids and comets.

And yet, it was precisely this chaos that laid the foundation for life. Those early collisions brought with them vital ingredients: water, carbon, nitrogen, and other life-giving elements, seeding the Earth with the materials needed for creation. Over time, Earth cooled, its surface hardened, and volcanic activity began releasing gases, slowly forming an atmosphere.

Earth's Protective Armour: Magnetosphere

In the early chaos of Earth's formation, gravity pulled heavier elements like iron and nickel toward the centre. This separation created a dense, solid inner core surrounded by a molten outer core. As this liquid metal churned around the solid core, it began to generate powerful electric currents from the natural movement of molten metal within the core itself.

Where there's electricity in motion, there's magnetism. These currents created Earth's geomagnetic field, a giant, invisible shield surrounding the planet, extending tens of thousands of kilometres into space. Just like the vast electron clouds orbiting the nucleus of an atom, this magnetic shield represents over 99% of Earth's true nature, while the physical ground we walk on, the part we see with our eyes, accounts for less than 1%.

This magnetic field is far more than a fascinating scientific fact. It's one of Earth's greatest protections. It deflects harmful solar winds and cosmic radiation, keeping our atmosphere from being stripped away like it was on Mars. Without this invisible armour, Earth would be a lifeless rock, scorched by solar energy and bombarded by space radiation.

Thanks to this shield, Earth could eventually hold onto its atmosphere, form oceans, and become the thriving, living planet we call home.

Beginning of Life on Earth

Then came the oceans.

As the planet cooled, water vapour in the atmosphere condensed and fell as rain for millions of years. Slowly, vast oceans pooled across Earth's surface. And in those dark, warm, mineral-rich waters, life found its first home.

The ocean was the perfect cradle for creation. It buffered temperature extremes. It shielded delicate molecules from harsh solar radiation.

It provided a stable, nutrient-rich environment filled with carbon, hydrogen, nitrogen, and trace minerals—the raw ingredients of organic chemistry.

Then somewhere in those ancient seas, the miracle began.

Simple molecules began to self-organise into more complex structures, lipid membranes, amino acids, and strands of RNA. These weren't just random accidents. They were nature's response to a deeper intelligence: the flow of energy through water, light, heat, and minerals.

Life didn't begin with a sudden spark. It began with a flow. A flow of electrons, of heat, of light, through molecules that could store and replicate information. These flows birthed the first primitive cells, tiny bubbles of organisation in a chaotic sea.

From that point forward, Earth's surface was no longer inert. It was alive. Not just alive, but self-organising, intelligent, and evolving.

Light, Electromagnetism, Water: The Trinity of Life

Life on Earth exists because three forces came together in perfect harmony: light, electromagnetism, and water. Light from the Sun provided the energy, fueling chemical reactions, guiding biological rhythms, and driving evolution from the very beginning.

Earth's electromagnetic field made Earth safe from harmful solar and cosmic radiation. It helped organise molecules, shape electrical communication in cells, and stabilise the energy flows that all living systems depend on.

The ocean became the medium of life. It carried minerals, allowed energy to move, and supported the first biochemical reactions. It was the perfect substrate for life to form, evolve, and thrive.

These three forces—light, electromagnetism, and water—aren't just universal markers of life. They are the biophysical trinity that shapes all life on Earth, including mammals and ultimately, you.

Evolution of Life on Earth

Let's now explore the key turning points in the journey of life on Earth, moments so powerful they reshaped biology forever and laid the foundation for everything we are today.

2.4 Billion Years Ago: The Great Oxygenation Event

In Earth's early days, the atmosphere was harsh, dominated by methane, carbon dioxide, and nitrogen. Life was anaerobic, meaning it thrived without oxygen. But something extraordinary happened when cyanobacteria (ancient microbes) learned to photosynthesise using sunlight.

They began releasing oxygen as a byproduct.

At first, this oxygen was toxic to most life, leading to the first major extinction event. But for those who adapted, oxygen became a superfuel. This moment, known as the Great Oxygenation Event, was biology's first revolution.

Oxygen didn't just change the air. It changed life's chemistry. With oxygen, the stage was set for a new kind of energy production: one that would birth the mitochondria.

The Evolution of Life on Earth

1.5 Billion Years Ago: The Endosymbiosis Event – Mitochondria Emerge

Imagine two ancient cells, one engulfing the other. But instead of digesting it, they struck a deal. The engulfed cell evolved into the mitochondrion, a tiny energy generator that transformed oxygen and nutrients into ATP, the currency of cellular energy. This was not a small deal. This was the biological equivalent of inventing the combustion engine.

This event is called endosymbiosis, and it marks the rise of complex life, organisms with multiple cells, specialised tissues, and eventually, conscious brains.

Mitochondria are not just power plants. They are ancient ancestors living within every one of your cells, except red blood cells, descendants of this billion-year-old pact.

Without them, none of the higher forms of life, including us, would exist.

1.0 Billion Years Ago: Chloroplasts Emerge – Plant Photosynthesis Begins

A second endosymbiotic event gave rise to chloroplasts, allowing ancient algae and plants to perform photosynthesis, converting sunlight, water, and carbon dioxide into glucose and oxygen.

This was Earth's solar panel moment. Light energy could now be stored as food, and the air became richer in oxygen. This is when biology learned to capture the Sun.

These photosynthetic organisms shaped the atmosphere, laid the groundwork for complex food webs, and formed the basis of the entire biosphere.

530 Million Years Ago: The Cambrian Explosion – Life Gets Creative

Out of seemingly nowhere, life exploded in diversity. In just 20-25 million years, a blink in geological time, 32 animal phyla appeared, including ancestors of insects, fish, molluscs, and vertebrates.

This was the biological Big Bang.

The rise in oxygen levels, thanks to photosynthesis, may have fueled this creativity. Higher oxygen meant more energy per cell, enabling complexity, mobility, and innovation.

Organisms now had eyes, limbs, gills, and brains. Nervous systems were born. Sensation and behaviour emerged. Life began to interact, perceive, and respond.

500 Million Years Ago: The Ozone Layer Forms – Haemoglobin Evolves

As photosynthetic life continued releasing oxygen, something else happened. O_3, or ozone, formed in the upper atmosphere. This ozone layer protected Earth's surface from the Sun's harsh UV rays, making land colonisation possible.

But life needed another upgrade to survive on land: a better way to carry oxygen. There came haemoglobin, a protein that allowed animals to store and transport large amounts of oxygen. This enabled higher metabolism, faster movement, and bigger brains.

The Evolution of Life on Earth

Haemoglobin was animal photosynthesis 1.0, a way to capture and circulate the energy oxygen held, supporting more dynamic forms of life.

With this innovation, life marched out of the sea and onto the land.

65 Million Years Ago: The K–T Event – Rise of Mammals and Modern Humans

A massive asteroid struck near what is now the Yucatán Peninsula in Mexico, triggering global wildfires, earthquakes, and a catastrophic disruption of the food chain. Dust and debris blocked sunlight, collapsing ecosystems that relied on photosynthesis. After dominating Earth's ecosystems for over 180 million years, the dinosaurs suddenly vanished.

Mammals, once small, nocturnal, and living in the shadows, seized the opportunity. Warm-blooded, agile, and adaptable, they thrived in the cooler, darker post-impact conditions. Over time, they diversified rapidly into many forms; some took to the trees, others the seas, and a few began to walk upright.

Among them were the early primates, and within that lineage emerged the hominins, our direct ancestors.

Somewhere in Africa, a branch of primates began to walk on two legs, which freed their hands for tool use, fire-building, and carrying food. Over time, bigger brains, social cooperation, and eventually language evolved.

From this lineage, modern humans would eventually emerge, beings capable not only of survival, but of awareness, creativity, and culture.

ATOMIC GLOW

1.2 Million Years Ago: Human Melanin Evolves

As early humans adapted to life on open savannas, they underwent a major transformation: they began to lose most of their body hair. In other mammals, melanin, the pigment responsible for absorbing light and protecting against UV, was primarily located in fur. But in hairless humans, melanin migrated to the skin, acting as both a solar panel and a shield.

This evolutionary upgrade was especially important as humans developed larger, more complex brains, specifically two highly evolved frontal lobes responsible for planning, language, and creativity. These new energy demands required more efficient ways to interface with light.

In this way, melanin became animal photosynthesis 2.0, an evolutionary leap that allowed humans not only to survive under the Sun, but to thrive in it, adapting to new environments across the globe while tuning their biology to the rhythm of light itself.

The Wisdom of Evolution

Earth is not just a planet; it's an intelligent, responsive system. A living library of adaptation. A home that evolves in rhythm with forces.

From the fiery birth of the planet to the gentle emergence of oceans, from the rise of mitochondria to the dance of dinosaurs, and finally to the awakening of human consciousness, evolution has always been guided by deeper laws, the science of nature.

Life is not random. It is powered by light, moved by electromagnetism, and carried by water. These three forces, light, electromagnetism, and water, are the fundamental elements that not only made life possible

The Evolution of Life on Earth

but also sustain it on this planet. They are the energetic threads weaving all biology together.

Nature does not force. It simply flows and evolves.

Just like electrons flow through atoms, rivers through valleys, or breath through lungs, life is an effortless, quantum dance of resonance and renewal. Every adaptation, every mutation, every leap forward was nature responding in perfect coherence with her environment.

The story of evolution is not just a history of life on Earth; it's a reflection of the deepest intelligence in the universe. And we, as humans, are not separate from it. We are cells within nature's body, born of her light, her water, and her magnetic field.

If we listen closely, this story teaches us something profound: that resistance is not the way of life. That coherence, flow, and alignment are the true engines of growth.

May we always remember this wisdom, written into our bones, whispered through our cells. The Earth knows. Life knows. And so do you.

You are the story of the Earth.

CHAPTER 9

THE ENGINES OF HUMAN LIFE

*"You are not the prisoner of your genes,
but the master of your energy."*
Alison Davis

For decades, we were taught that our genes were our blueprint—unchangeable, fixed, and fated. Nuclear DNA was seen as the master code, locked within us, determining everything from the colour of our eyes to whether we'd age like fine wine or a bruised banana. The narrative was simple: you are what your genes say you are.

Epigenetics: The End of the Gene Theory

Then came the Human Genome Project, completed in 2003. Scientists raced to map every gene in the human body, expecting to find hundreds of thousands of them. After all, we're incredibly complex creatures, right?

Instead, they made a jaw-dropping discovery: we have fewer than 25,000 genes. That's only slightly more than a roundworm. Yes, a worm. To add to the identity crisis, it turns out we share over 98% of our genes with chimpanzees. Yet look around, humans not only look and live differently, but we build rocket ships, write poetry, file taxes, and suffer from existential dread. Chimps? Not so much.

So if genes are supposed to be the master key to life, how is it that our lives are so radically different, despite being nearly identical to chimps at the genetic level?

The answer cracked open a whole new field of science: epigenetics.

Epigenetics flipped the script. It showed that genes are not our destiny; they are options. Think of your DNA like a symphony orchestra. Every gene is an instrument. But without a conductor, there's no music, just noise. The conductor is your environment: light, water, temperature, food, movement, thoughts, emotions, electromagnetic fields…all of it.

These cues tell your genes what to express, when to turn on or off, and how loudly to play. Your DNA isn't static; it's responsive. It listens to your environment and acts accordingly. Suddenly, the age-old belief that you're doomed by your family history is obsolete.

We are not victims of our nuclear DNA. We are the authors of our biology. And as mind-blowing as this sounds, this was just the beginning.

The Engines of Human Life

Mitochondria: A New DNA

Remember the rise of mitochondria through the ancient endosymbiosis event we explored earlier? These masterful engines of energy, originally free-living bacteria, joined forces with early cells and have been working silently in the background of the evolution of life on Earth for over 1.5 billion years.

Fortunately, their silence didn't last forever.

In the late 1970s, a groundbreaking shift occurred. Dr. Doug Wallace, a visionary in mitochondrial biology, turned his attention to these humble powerhouses. What he discovered changed everything.

Mitochondria, it turns out, have their own DNA, separate from the nuclear DNA that makes up most of our genetic code.

And here's the twist: this mitochondrial DNA (mtDNA) doesn't come from your father. It's passed down only through your mother, an unbroken maternal lineage stretching back through time, like a biological memory encoded in your very cells.

While the nuclear genome contains around 23,000 genes, mtDNA holds just 37. But don't be fooled by the number. These 37 genes play a monumental role in your health and vitality. Dr. Wallace's research revealed that dysfunction in these mitochondrial genes is at the root of nearly every modern chronic illness, diabetes, cancer, Alzheimer's, autoimmune diseases, and even the ageing process itself.

The more we study them, the clearer it becomes: mitochondria are not just background players in the story of life. They are the directors of the show.

Where Do Mitochondria Live?

Mitochondria are found in nearly every cell of the human body, except red blood cells. Each cell contains anywhere from hundreds to thousands of these energy-producing organelles, depending on how much energy that cell needs.

Cells with high energy demand, such as those in the brain, heart, and gut lining, contain the highest concentrations of mitochondria. For instance, heart muscle cells can hold up to 5,000 mitochondria per cell to keep the heart beating continuously. The brain, though only 2% of our body weight, consumes about 20% of our total energy, reflecting the dense mitochondrial presence in neurons. Likewise, the cells of the gut lining undergo rapid turnover and require enormous amounts of energy to continuously renew and maintain the barrier that supports digestion and overall health.

This distribution explains why chronic illnesses most often strike the brain (neurological disorders), the heart (cardiovascular disease), and the immune system (autoimmune diseases). These are not random; they are mitochondrial diseases at the root.

- Cardiovascular disease remains the leading cause of death globally, responsible for an estimated 17.9 million deaths each year.
- Neurological diseases like Alzheimer's and Parkinson's are rising sharply, with Alzheimer's affecting over 55 million people worldwide.
- Autoimmune conditions like lupus, rheumatoid arthritis, and Hashimoto's have tripled in incidence over the last few decades.

Mitochondrial dysfunction underlies them all. Modern illnesses are not due to faulty genes, but disrupted energy production in cells that need it most.

Roles of Mitochondria

Mitochondria are best known as the "powerhouses" of the cell, the engines of life. Just like a car depends on its engine for performance, reliability, and longevity, your body depends on mitochondria for its health, vitality, and resilience. They are at the very core of how well you function, feel, and even age.

Converting Food into Energy

Through an elegant process called oxidative phosphorylation, mitochondria take the oxygen you breathe and the nutrients from the food you eat and turn them into ATP, the universal "energy currency" of life.

Here is how it works at a high level: once you eat, your digestive system chops food down into smaller molecules: carbohydrates become glucose, fats become fatty acids, and proteins become amino acids. These smaller molecules then enter a central energy hub called the Krebs cycle (or citric acid cycle), where hydrogen atoms are stripped away from these molecules and broken down further, until they are separated into their core components: electrons and protons. Yes, at its deepest level, food is ultimately broken down into tiny subatomic particles!

From there, the electrons embark on a journey through the Electron Transport Chain (ETC), a series of protein complexes (I through IV) embedded in the inner mitochondrial membrane. As electrons move through the chain, they release energy, which is used to pump protons out of the mitochondrial matrix and into the intermembrane space. This builds up a powerful proton gradient. Finally, these protons rush back in through Complex V (ATP synthase), spinning the ATP synthase like a tiny turbine.

This spinning generates ATP, which fuels everything you do. From the blink of an eye to the beat of your heart, from breathing and thinking to growing your hair and nails, running marathons, digesting food, and even forming memories, none of it would be possible without mitochondria. It also powers virtually every cellular function inside your body, from repair and regeneration to communication and defence.

Moment by moment, mitochondria are what keep you alive. Without adequate cellular energy, the body simply cannot function optimally, regardless of how healthy your diet, lifestyle, or mindset might be.

The Command Centre

Mitochondria are far more than cellular "power plants." They are the true command centres of health, performance, and longevity. They decide whether cells live or die, regulate inflammation, orchestrate calcium signalling, and even influence which genes are turned on or off.

What makes them even more extraordinary is how dynamic they are. Mitochondria are not fixed machines; they are highly responsive, constantly adapting to the environment you live in. They can be strengthened or weakened depending on the signals they receive. Those signals come from every part of your daily life, from the visible light of the morning sun to the invisible electromagnetic fields around you, from the food you eat to the very thoughts you think.

Every sunrise you watch, every late night spent under artificial light, every deep breath you take, every stressful thought, all of it shapes how your mitochondria function. The quality of these signals directly impacts the quality of their decisions, which, in turn, impacts every part of your biology. We'll explore this in greater depth in the next chapters.

Mitochondria are the bridge between your environment and your biology. Every choice you make is either upgrading your mitochondria or wearing them down. When you nurture them, they don't just make energy; they upgrade you. But when they begin to fail, the consequences ripple across the entire body. Cells struggle to function. Organs weaken. Ageing accelerates. And disease inevitably follows.

The Ultimate Engine of Life

In summary, mitochondria house an engine unlike anything humanity has ever built, an engine that not only produces energy with unparalleled precision to power your cells but also orchestrates the harmony of your entire internal ecosystem.

Mitochondrial health isn't just important to biology; it's the core of human vitality. When your mitochondria thrive, you thrive, physically, mentally, emotionally, and even spiritually. This is where true health begins.

Heteroplasmy of Mitochondria

Let's talk about heteroplasmy, a term that sounds complex but holds the key to understanding how and why mitochondrial function breaks down.

In simple terms, heteroplasmy refers to the presence of both healthy and damaged mitochondrial DNA (mtDNA) within the same cell. While we all carry a mix, it's the ratio that matters. The higher the percentage of mutated mtDNA, the greater the risk of dysfunction.

Naturally, heteroplasmy increases with age. Starting around age 30, the proportion of damaged mitochondrial DNA tends to rise by about 10% each decade. This slow and steady accumulation is one of the main biological markers of ageing.

But ageing isn't the only driver. A child can also inherit a high heteroplasmy rate from their mother, especially if the mother lived in a toxic or stressful environment that disrupted her mitochondrial health. When this happens, the baby is born with a cellular disadvantage. If the damaged mitochondria are not renewed or repaired, the dysfunction can carry on for multiple generations.

This is why we're seeing an alarming rise in adult-onset diseases in children, things like type 2 diabetes, autoimmune conditions, obesity, and neurodevelopmental disorders. These children aren't biologically old in years, but their mitochondria are ageing prematurely.

Understanding heteroplasmy helps us reframe disease and ageing, not as a mysterious breakdown of the body, but as a clear signal that our cellular engines are in distress.

Mitochondria Regeneration

Yes, you can upgrade, strengthen, and even regenerate your mitochondrial colonies. That's one of the most empowering truths in biology, especially if you're someone with a high heteroplasmy rate, whether inherited or acquired over time.

The Engines of Human Life

Let's say you were born with mitochondrial DNA that carries what scientists call a high heteroplasmy rate. You might think you're stuck with weak cellular engines for life. But that's far from the truth.

Your body is brilliantly designed. It has the built-in ability to create new mitochondria through a process called mitochondrial biogenesis, literally the birth of new mitochondria. At the same time, it can remove damaged or dysfunctional ones through a process called mitophagy, which is like cellular spring cleaning. During mitophagy, your cells identify malfunctioning mitochondria and safely dismantle and recycle them, preventing them from causing further harm.

They are natural, intelligent systems your body is always ready to activate when given the right signals. You are not stuck with bad engines. You can reboot, rebuild, and recode your entire mitochondrial network.

In the chapters ahead, we'll explore exactly how to do that. You'll discover the lifestyle choices, environmental cues, and mental frameworks that encourage your body to repair and renew its energetic infrastructure. You'll learn how to work with nature, not against it, how light, water, and electromagnetism can trigger the renewal of your cellular power plants.

Once you understand mitochondria for what they really are—your biological engines, your energetic core—everything shifts.

You'll realise that health is not a mystery. It's not luck. It's not genetics. It's physics. It's energy. It's coherence.

And you hold the key.

CHAPTER 10

OPTIMAL INPUTS: FUELS FOR THE PERFECT ENGINE

"You are nature, illuminated from within."
Alison Davis

As the saying goes, quality inputs create quality results. This principle isn't just true in business, creative projects, or gardening; it's also at the heart of your biology.

Your mitochondria, the quantum engines of life, thrive when they're fueled with the right inputs. Feed them well, you'll be rewarded with vibrant energy, mental clarity, emotional resilience, and a glow that turns heads. Feed them junk, and they'll start throwing tantrums in the form of fatigue, inflammation, brain fog, and eventually chronic disease.

In this chapter, we'll examine what essential inputs your mitochondria truly need to run optimally. Let's just say, they don't run on willpower or coffee (although we've all tried), but the fundamental forces of nature they've evolved to recognise and thrive on.

We begin with the most potent one of all: sunlight.

Essential Input One: Sunlight – The Original Nutrient

As explored in earlier chapters, light is not merely a backdrop to life; it is the activator of the universe itself. For life on Earth, sunlight is the spark behind photosynthesis, the driver of circadian rhythms, the calibrator of hormones, and the initiator of cellular energy. Without sunlight, life as we know it would not exist. Your mitochondria are no exception.

Mitochondria are exquisitely attuned to sunlight. Far from being simple fuel burners, they are highly responsive organelles and biological sensors that interpret light cues to regulate energy production, initiate repair pathways, and optimise cellular function. In this sense, sunlight isn't just something you see, it's something you are.

Your body is fundamentally designed to interact with sunlight. You don't merely witness it through your eyes, you absorb it, respond to it, and even metabolise it.

Wavelengths of Sunlight

Here's how the different wavelengths of sunlight influence your mitochondria:

Optimal Inputs: Fuels for the Perfect Engine

UVA Light (Sunrise to Sunset. Invisible. 320–400 nm)

UVA light, which makes up about 95% of the UV radiation that reaches Earth's surface, penetrates deeply into the skin and initiates the release of nitric oxide (NO), a molecule that relaxes blood vessels and enhances circulation. NO also temporarily slows mitochondrial respiration. This acts as a form of cellular "maintenance mode," reducing oxidative stress and metabolic demand. It's also why sunlight often suppresses appetite; your mitochondria are resting and don't need extra fuel.

UVB Light (Later Morning to Early Afternoon. Invisible. 280–320 nm)

UVB light is more intense but shorter in wavelength. While it doesn't penetrate as deeply, it plays a vital role in converting cholesterol in your skin into vitamin D. Vitamin D is no minor player; it helps stabilise mitochondrial membranes, regulate intracellular calcium (key for signalling and energy transfer), and modulate immune and neurological function. In short, UVB light upgrades mitochondrial stability from the inside out.

Visible Light (Sunrise to Sunset. Visible. 400–700 nm)

Visible light is the spectrum your eyes can see, the rainbow spanning from violet to red. It's the primary driver of circadian rhythm, as specialised photoreceptors in your eyes send signals to your brain's master clock (the suprachiasmatic nucleus). This synchronises your entire body with the day–night cycle.

Within cells, visible light penetrates tissues and interacts with chromophores inside mitochondria. Each wavelength carries unique effects: for example, blue light in the morning strengthens the circadian rhythm and sharpens alertness, while red light penetrates more deeply into tissues, enhancing ATP production, regulating reactive oxygen species (ROS), and supporting mitochondrial health, tissue repair, and resilience.

Near-Infrared Light (NIR) (All day. Invisible. 700–1400 nm)

NIR penetrates deeply into the body, even reaching muscles, joints, and internal organs. At the mitochondrial level, NIR interacts with cytochrome c oxidase (complex IV) in the electron transport chain, enhancing ATP production, improving oxygen utilisation, and reducing oxidative stress.

NIR also helps structure water inside cells, strengthening the exclusion zone (EZ water) that supports cellular communication and energy efficiency. In short, NIR is like a cellular battery charger, energising tissues while lowering inflammation.

Infrared Light (IR) (All Day. Invisible. 1400 nm–1 mm)

IR makes up 47% of the Sun's energy that reaches Earth, and is mostly experienced as heat. Unlike NIR, which acts more directly on mitochondria, IR interacts strongly with water molecules in the body, increasing the structured water layer (EZ water) around proteins and membranes. This improves hydration, nutrient flow, and waste removal at the cellular level.

Infrared also promotes vasodilation (widening of blood vessels), improving circulation and oxygen delivery. Its gentle heating effect relaxes muscles, lowers stress, and enhances detoxification.

Optimal Inputs: Fuels for the Perfect Engine

Harnessing the Sun

There are three primary pathways through which sunlight enters the body and influences your mitochondria:

1. Skin: The Solar Panel

Your skin isn't just a protective barrier; it's your body's largest light-sensing organ and a vital contributor to mitochondrial health.

Each time sunlight touches your skin, especially in the early morning, it initiates a symphony of biological processes that go far beyond simply tanning.

Think of your skin as a solar panel. Just as solar panels convert sunlight into electricity, your skin converts different wavelengths of sunlight into cellular signals. These signals regulate healing, energy production, and hormonal balance, supporting your body's natural rhythm and vitality.

2. Eyes: The Time Master

Your eyes are more than windows to the soul; they're the command centre for your body's internal clock.

Inside your retina are specialised light-sensing cells called intrinsically photosensitive retinal ganglion cells (ipRGCs). These cells aren't responsible for vision, but for light detection. They send signals directly to your suprachiasmatic nucleus (SCN), a tiny region in your brain that acts as the master clock.

This master clock governs nearly every biological rhythm in your body, your hormone production, body temperature, sleep-wake cycles, metabolism, and yes, even how well your mitochondria produce energy.

Morning sunlight matters most. It contains the full spectrum of wavelengths, including blue light, which tells your brain, "It's daytime. Wake up and be alert." This sets your circadian rhythm in motion for the day and signals mitochondria to ramp up energy production.

3. Food: Barcode of the Sun

Food is, in essence, condensed sunlight. Through photosynthesis, plants capture photons from the Sun and lock that energy into chemical bonds, creating sugars, fats, proteins, and other compounds. This is how solar energy is translated into matter.

When you eat plants or an animal that ate plants, you're consuming that stored solar energy held in the electrons of chemical bonds. During digestion and the Krebs cycle, those bonds are broken down, releasing electrons that are ultimately delivered into your mitochondria, the engines that turn sunlight's gift into ATP, the energy of life.

The quality of light a food is exposed to imprints its quantum signature. A tomato ripened under full-spectrum sunlight in local soil carries a very different photonic and nutrient profile than one grown under artificial light in a distant greenhouse. The first delivers coherent information to your mitochondria, reinforcing natural rhythms and biological harmony. The second introduces mismatch, a distortion between light, energy, and biology that the body struggles to interpret.

That's why choosing food that's in-season, locally grown, and ripened under real sunlight isn't just a nutritional decision, it's a quantum one. Each bite delivers photonic information your mitochondria were designed to decode. The more coherent and aligned that light information is, the more clearly your biology can function, heal, and thrive.

Optimal Inputs: Fuels for the Perfect Engine

Skin & Eyes: The Home of Critical Master Regulators

Your skin and eyes aren't just passive receivers of light; they're home to some of the most powerful regulators in your hormonal and biological pathways.

Leptin: The Master Metabolic Switch

Leptin is a hormone stored in your fat cells, and your skin contains plenty of those. It acts as a master signal to your brain (specifically the hypothalamus), constantly reporting on your body's energy status, whether you've got enough fuel or need to conserve energy. It is the energy accountant of your body.

Leptin is intricately tied to your circadian rhythm, and its communication depends heavily on natural light. Morning sunlight on your skin and retina helps synchronise your internal clock, which in turn helps regulate leptin's release.

If you're missing morning light and bathing in artificial blue light at night, you're likely scrambling this crucial signalling. The result could be weight gain, fatigue, poor sleep, and even accelerated ageing. Because when leptin can't talk to the brain, mitochondria don't get the right instructions. And confused mitochondria? Let's just say they don't exactly power your glow.

Melanin: The Advanced Solar Panel (and Battery)

Melanin is best known as the pigment that gives colour to your skin, hair, and eyes, and as a shield that protects you from harmful UV rays. But melanin is far more than just a pigment or a biological sunscreen; it is a dynamic molecule with powerful roles in light absorption, energy transfer, and cellular protection.

Melanin functions as a light transducer, an advanced solar panel built into your body. It has the extraordinary capacity to absorb nearly the entire electromagnetic spectrum, from low-frequency radio waves all the way to high-energy gamma rays. Even more remarkably, melanin can transform this absorbed energy into other usable forms, including heat and free electrons that support cellular energy processes. In doing so, it bypasses the mitochondria altogether, offering an alternate pathway for energy generation.

Emerging research suggests that melanin may also store solar energy and release it later, essentially acting as a biological battery and an active energy processor. It's not just shielding you, it's supporting your energy systems in ways science is only beginning to uncover.

Serotonin & Melatonin: Day-Night Biochemical Twins

Serotonin and melatonin are often seen as separate players, one linked to happiness, the other to sleep. But they're actually two sides of the same coin, shaped by your relationship with natural light.

Serotonin production is heavily influenced by sunlight exposure, especially through the eyes and on the skin in the early part of the day. It's your "feel-good" neurotransmitter, helping regulate mood, appetite, motivation, and overall daytime energy. This is why getting outside in the morning can feel like a natural antidepressant; it literally is.

But serotonin's role doesn't end at sundown. As darkness falls, your brain converts serotonin into melatonin, the hormone that governs deep, restorative sleep and overnight mitochondrial repair. Without sufficient morning sunlight to kickstart serotonin, your melatonin levels will be compromised later in the evening. When melatonin is low, so is your ability to recover and regenerate.

Optimal Inputs: Fuels for the Perfect Engine

Melatonin is often misunderstood as just a sleep aid. In reality, it's one of your body's most powerful antioxidants and mitochondrial protectors. It directs cellular clean-up crews, regulates autophagy (your cells' recycling process), and helps orchestrate DNA repair and immune function while you sleep. It's like your body's night-shift supervisor, ensuring everything gets restored, detoxified, and prepped for another day of energy production.

Essential Input Two: Earth's Electromagnetic Field

Just as a plant's roots rely on the Earth's electromagnetic field to orient and grow, mitochondria depend on this same invisible force to function.

Mitochondrial Membrane Stabiliser

Mitochondria aren't just little engines churning out energy; they're quantum sensors. Their membranes are finely tuned antennas, constantly picking up signals from the environment. One of the most important signals is the Earth's electromagnetic field.

Each mitochondrion maintains a voltage across its inner membrane of about −150 to −180 millivolts, which translates into an electric field strength of roughly 30 million volts per meter. Think of this voltage like the tension in a slingshot: the greater the tension, the more powerful the release. In this case, the "release" is ATP, your body's main energy currency.

But this mitochondrial voltage doesn't arise on its own; it relies on exposure to natural electromagnetic inputs, including the Earth's field, to stay charged and functional. The Earth's geomagnetic field plays a direct role in maintaining the membrane potential of mitochondria by stabilising electron flow along the electron transport chain, the final stage of ATP production. This process is exquisitely sensitive to electromagnetic input. Too little, the chain becomes inefficient or even stalls. Too much, for example, from man-made EMFs, it can become chaotic or misdirected.

Your Cellular Tuner

Every living cell is designed to operate in a symphony with the planet's rhythms. The Earth pulses with a steady frequency, most notably the Schumann Resonance at around 7.83 Hz. It closely mirrors the alpha brain waves in your brain, which are linked with creativity, calm focus, healing, and deep rest.

This isn't a coincidence. It's coherence.

Your brain, nervous system, and mitochondria are looking to sync to this natural rhythm. Without regular exposure to these frequencies, your internal systems start to lose coherence, leading to fatigue, inflammation, poor sleep, and slower recovery.

The Ultimate Human Charger

All life on Earth is designed to live in direct connection with the planet and this field. Trees are rooted in it. Animals walk on it. Birds perch in trees and ride the currents of the field through the air. Humans, uniquely, are the only mammals with sweat glands on the palms of our hands and the soles of our feet. That's not random design. It's biology's invitation to connect and discharge, to ground ourselves to Earth.

When your bare skin touches the Earth, whether it's soil, grass, rock, sand, or ocean water, you literally plug into the planet's electromagnetic field. You become a living conduit, syncing with the natural frequency of the Earth.

Optimal Inputs: Fuels for the Perfect Engine

Bathing in the Earth's field doesn't just feel good, it:

- Increases ATP production, meaning more cellular energy for your brain, muscles, metabolism, and repair.
- Enhances cellular communication and coherence, so your cells can signal, repair, and regenerate with precision.
- Reinforces your circadian rhythm, the master clock that mitochondria depend on to function efficiently.
- Supports deep healing, especially during sleep, when regeneration peaks.

In today's world, dominated by chaotic artificial EMFs, from Wi-Fi to Bluetooth to 5G, the Earth's natural field, a stabilising, coherent healing force, is needed more than ever.

Essential Input Three: Oxygen

The Great Oxygenation Event 2.4 billion years ago made complex life possible. Oxygen has become the superfuel that enabled the explosion of multicellular life. But oxygen is a powerful force, both life-giving and potentially destructive. So nature had to evolve an elegant system to deliver it safely and precisely.

Haemoglobin: The Oxygen Deliverer

About 500 million years ago, as life evolved from simple to complex forms, haemoglobin emerged, a protein perfectly designed to carry oxygen through the bloodstream.

Each haemoglobin molecule contains iron at its core, and here's where it gets quantum: iron must be in the +2 oxidation state (Fe^{2+}) to bind oxygen. If it oxidises to +3 (Fe^{3+}), oxygen can no longer dock. In that state, it becomes methemoglobin, and oxygen delivery shuts down.

This detail might seem tiny, but it's crucial. Life depends on maintaining the delicate balance of oxidation, and nature responded to this challenge by forming the ozone layer, which filters out the high-energy ultraviolet radiation that would otherwise oxidise Fe^{2+} to Fe^{3+}. Without the ozone layer, complex life couldn't exist on land.

Where Does Oxygen Go?

Haemoglobin carries oxygen from your lungs and delivers it to where it's needed most, your brain, muscles, organs, and all the hardworking tissues that keep you alive and functioning. These areas have high energy demands and rely on a constant oxygen supply to operate at their best.

Once delivered, oxygen doesn't just float around aimlessly. It's shuttled deep into your tissues, into every single cell, and ultimately into the mitochondria, your body's energy generators.

Mitochondria: Oxygen's Final Destination

At the tail end of the electron transport chain, oxygen plays its starring role as the terminal electron acceptor. This is where all the electrons, stripped from your food, are finally passed to oxygen.

Why here? Because oxygen is electronegative, meaning it attracts electrons strongly. By placing it at the end of the chain, nature ensures that electrons flow in a single direction, like a waterfall. This flow powers the creation of a proton gradient, which drives the turbine-like ATP synthase to spin and generate ATP, your cellular energy currency. Without oxygen at the endpoint, this entire chain grinds to a halt, and energy production collapses.

Optimal Inputs: Fuels for the Perfect Engine

Oxygen isn't just the air we breathe; it's the vital spark that ignites life itself.

From the miraculous birth of haemoglobin to the elegant design of the electron transport chain, nature has engineered a masterpiece, a system where oxygen takes centre stage as the ultimate receiver, completing the energetic dance within our cells.

Without oxygen, energy halts. Life pauses. Nothing flows.

As you will discover in the next chapter, Oxygen is not just crucial for energy production; it is also irreplaceable for the production of two other vital outputs, structured cellular water and biophotons.

Essential Input Four: Water

Every cell in our body depends on water, and we can't survive without it for more than a few days. While we do get some water from food, especially fresh fruits and vegetables, we also rely heavily on the water we drink. Since we've already explored food in the Sunlight section, here we'll focus specifically on the water we drink.

Mitochondrial Supporter

Water's most well-known role is hydration, but this is often misunderstood. True hydration, however, happens inside our cells, where mitochondria create a special form of metabolic water. We'll explore this deeper in the next chapter.

Drinking water plays a crucial supporting role in mitochondrial function. It:

- Maintains blood plasma fluidity, enabling oxygen and nutrient delivery to cells, which is essential for mitochondrial respiration.
- Regulates body temperature, waste removal, and pH balance, all of which influence mitochondrial health and efficiency.
- Supports the hydration shell around proteins, helping enzymes (including those in the electron transport chain) fold and function properly.

Quality of Drinking Water

It's not just about quantity; it's also about the quality and structure of the water we drink.

Natural, clean, mineral-rich water, like spring water, closely mimics the metabolic water mitochondria produce. This lowers the energetic cost for mitochondria and helps maintain cellular coherence.

Water that's structured, mineralised, and free from contaminants could support mitochondrial function better than processed, demineralised, or chemically treated tap water.

Our Generous Life-giving Mother

When mitochondria are well-nourished by sunlight, Earth's electromagnetic field, oxygen, and natural water, they operate with graceful precision. They continually renew themselves through mitobiogenesis, keeping their community of mitochondria healthy and balanced. This helps maintain a low heteroplasmy rate (the ratio of damaged to healthy mitochondria), ensuring they produce clean, abundant energy along with the other vital outputs your body needs to thrive.

Optimal Inputs: Fuels for the Perfect Engine

When mitochondria thrive, so do your most mitochondria-dense, energy-demanding organs: the brain, heart, and gut. This means sharper cognition, mental clarity, emotional stability, optimal heart rate variability, stronger digestion, and better nutrient absorption. In short, when your mitochondria are aligned with nature's gifts, your entire being moves toward vitality and resilience.

Mother Nature provides all of these freely and generously. She is the original source of nourishment and healing, offering us light to awaken our cells, water that remembers life, food that carries the Sun's intelligence, air charged with oxygen to fuel our breath, and Earth's gentle magnetic rhythms to ground and restore us.

We don't need to earn it. All we need is to return to her, our Mother, our home.

To step barefoot on the soil. To greet the morning sun with open eyes. To eat what grows under the sky. To breathe deeply under trees. And to remember, we are not just of nature, we are nature.

CHAPTER 11

VITAL OUTPUTS: THE POWER YOU BECOME

"The greatest power is realising you are the power."
Alison Davis

Now that you've delivered the perfect fuels—sunlight, natural water, Earth's natural field, and abundant oxygen—to your miraculous quantum engines, it's time to witness the brilliance of what your mitochondria do with them.

As we explored earlier in Chapter 9, ATP, the body's energy currency, is one of the most fundamental outputs of your mitochondria. But ATP is just the beginning. Your mitochondria are conscious transducers and master alchemists of nature, transforming raw materials into five other vital outputs that go far beyond energy production, outputs that shape not just your health, but who you become.

What you're about to discover may forever change how you view vitality and the language of life. Let's begin with one of the most astonishing revelations: Ultraweak Photon Emissions, the subtle light as a signature of health, consciousness, and coherence.

Power Output One: Ultraweak Photon Emissions

Every living cell in your body is quietly shimmering with light. Not the kind of light you can see with your eyes, but an unimaginably faint glow, so delicate it takes specialised equipment to detect it. This phenomenon is called Ultraweak Photon Emissions (UPE), also known as biophotons.

Discovery of UPE

The discovery dates back to the early 20th century, when Russian scientist Alexander Gurwitsch observed that onion roots exposed to ultraviolet light began to grow faster than other onion roots. He proposed that cells communicate using light, a bold idea for its time. Later, in the 1970s, German physicist Fritz-Albert Popp developed ultra-sensitive photomultiplier devices that confirmed every living organism emits this subtle light. His work proved that UPE isn't just a curiosity; it's a fundamental aspect of biology.

How Mitochondria Produce UPEs

Mitochondria are, by far, the brightest beacons of light within your cells. While red blood cells lack mitochondria and produce UPEs via oxidative haemoglobin chemistry, in all other cells, mitochondria are the primary source of UPEs.

Vital Outputs: The Power You Become

During oxidative phosphorylation, while generating ATP, mitochondria produce a normal byproduct called reactive oxygen species (ROS), like superoxide and hydrogen peroxide. When these excited molecules relax, they release tiny packets of light, biophotons. These photons are coherent (laser-like) and can interact with other cells and tissues, forming part of the body's electromagnetic signature.

Researchers such as Roeland van Wijk have documented these links between mitochondrial respiration, ROS production, and measurable UPE, and have proposed that careful analysis of UPE patterns could serve as indicators of mitochondrial health and stress.

The Spectrum of Mitochondria UPEs

According to neurosurgeon and quantum biologist Dr. Jack Kruse, healthy mitochondria emit UPEs predominantly in the 100–400 nm range. This lies mostly in the ultraviolet spectrum, including wavelengths that never reach Earth's surface from the Sun. For example, UVB from sunlight begins around 280 nm, and UVC (100–280 nm) is completely filtered out by the Earth's atmosphere and magnetosphere.

This means your mitochondria are not simply reflecting sunlight; they are transforming it. They take in nature's energy and re-emit it as more powerful, coherent, biologically tuned photons in a spectrum your body cannot receive from the outside world.

Research also shows that in less optimal health states, the UPE spectrum broadens, often extending into the visible and near-infrared range (up to ~800 nm or more). This shift can indicate reduced coherence and higher oxidative stress, meaning the light your cells emit is less organised and less efficient in carrying biological information.

ATOMIC GLOW

Melanin: The UPE Enhancer

As we explored in earlier chapters, melanin, the pigment that gives skin, hair, and eyes their colour, is far more than a sunscreen. It is a broadband solar absorber, capable of capturing the full spectrum of sunlight, from radio frequencies, visible spectrum, all the way to gamma rays.

But melanin is not confined to your skin. It's also present inside many tissues and cells in your brain, eyes, and ears, with much of it strategically positioned near mitochondria. This close proximity is no accident; it's part of a highly evolved photonic partnership.

Melanin absorbs a wide range of solar frequencies into its complex, fractal-like molecular structure. When this energy interacts with water and oxygen, melanin can split water molecules, producing both ROS and UPEs. These photons are then available within the cell to influence mitochondrial function.

In this way, melanin acts both as an energy converter and a communicator of light. It absorbs sunlight and transforms it into a local field of energy-rich, information-packed light that can directly influence mitochondrial DNA activity. As a result, mitochondria emit UPEs that are more coherent and often concentrated in the ultraviolet range of about 200–400 nm. This finely tuned, highly ordered light is believed to help synchronise cellular processes and improve the precision of biological signalling.

Why UPEs Matter

UPEs are far more than a faint cellular glow; they are the language of life. These photons are carriers of information, transmitting signals between cells at the speed of light and orchestrating processes from

Vital Outputs: The Power You Become

enzyme activation to DNA repair. The coherence or order of this light reflects the coherence of your biology itself.

When your mitochondria are well-fueled and in sync with nature's perfect inputs, this biophotonic communication becomes structured, precise, and efficient. Think of UPE as your body's internal fibre-optic network: when the signal is crisp, every cell is in harmony; when it's scrambled by environmental stressors (we'll get to those in the next chapter), the messages distort and your biology begins to unravel.

This light is also a marker of vitality. High coherence in biophoton emissions has been observed in seasoned meditators and people living in close rhythm with the natural world. The brighter and more ordered your inner light, the more powerfully you can express life itself.

On a biochemical level, some of your most important signalling molecules depend on UPEs. Leptin, your master energy accountant, has a peak absorption near 220 nm. Melatonin, serotonin, dopamine, and other neurochemical regulators also operate within this range. Without the precise wavelengths of UPE to trigger them, these molecules cannot perform as designed, leading to cascading breakdowns in cellular pathways.

Your UPE is not a byproduct; it's a master signal. It is both a mirror and a driver of health, linking the quantum nature of light to the very fabric of your physiology.

UPEs' Role in Forming the Biofield

UPEs emitted from the mitochondria and blood don't remain confined. They can propagate through structured water networks, connective tissue (fascia), and the extracellular matrix, extending beyond the body's surface. These photons can be visualised in certain experimental setups, such as Kirlian photography.

When combined with ionic currents (flows of charged particles) from the nervous and cardiovascular systems, they contribute to the body's surrounding electromagnetic field, often referred to as the biofield.

Power Output Two: Bioelectromagnetic Field

If UPE is your body's fibre-optic light network, then your bioelectromagnetic field (The Field) is its invisible broadcast signal, an energy field generated directly from the dance of electrons and protons inside your mitochondria.

How the Field Is Generated

Every time your mitochondria produce ATP through oxidative phosphorylation, charged particles, electrons, and protons flow through the electron transport chain (ETC) embedded in the inner mitochondrial membrane.

Where there's electricity in motion, there's magnetism. Moving charges create electromagnetic fields. At the nanoscale, this happens trillions of times per second in every cell. The result is a faint, yet highly ordered, electromagnetic field that surrounds and penetrates your tissues.

Reactive oxygen species (ROS), the byproducts of mitochondrial respiration, also contribute to this field. When ROS interact with other molecules, they produce tiny fluctuations in electrical charge and release photons, both of which can alter the local electromagnetic environment. These micro-events collectively shape the texture and frequency profile of your bioelectromagnetic field.

Vital Outputs: The Power You Become

How the Field Shapes Your Internal Architecture

Biophotons, those ultraweak pulses of light, are part of the electromagnetic spectrum, meaning they can both generate and be influenced by electromagnetic fields. Just as electrons orbiting an atom are shaped by the fields they move within, the coherence and quality of your biophotons are tuned by the bioelectromagnetic field produced by your mitochondria.

This is not a passive occurrence but part of a highly sophisticated feedback system. Your bioelectromagnetic field influences the movement of charged particles inside your cells, which in turn affects biochemical reactions, mitochondrial communication, and even the structuring of water within cells. The more coherent your field, the more precisely your internal processes function.

In the fetus, where mitochondria are densely packed in rapidly dividing cells during neurulation, these localised magnetic fields could create a "microenvironment" for signalling. This enhances the coherence of UPEs and supports highly precise developmental outcomes.

Melanin adds another layer to this tuning process. Acting like a biological antenna, melanin absorbs and redistributes electromagnetic energy, subtly reshaping the electromagnetic landscape inside the cell. This helps synchronise mitochondrial activity and fine-tune the spectrum of UPEs they emit.

Your bioelectromagnetic field is therefore not merely a byproduct; it is a real-time, dynamic architecture, constantly interacting with the quantum events in your mitochondria. It is light and field, wave and particle, continually bringing your biology into resonance with itself and the greater electromagnetic environment of Earth.

The Biofield & Spirituality: Your External Architecture

Just as all living organisms emit a halo of biophotons around their bodies, capturable in techniques like Kirlian photography, your bioelectromagnetic field extends well beyond your skin, also known as the biofield. Every cell contributes to it, but your brain and especially your heart dominate this field.

The Brain's Biofield

The brain generates an electromagnetic field primarily through synchronised firing of neurons and mitochondrial electron flow. This activity is detectable using EEG (electroencephalography) for electrical signals, and MEG (magnetoencephalography) for magnetic ones. The brain's field is primarily electrical, shaping thought, perception, and conscious awareness.

This biofield may help support the subconscious mind as a deeper interface with the quantum field, an infinite reservoir of patterns and possibilities beyond everyday awareness.

The Heart's Biofield

The heart, with the highest density of mitochondria in the body, produces an electromagnetic field that radiates in every direction. Sensitive instruments like magnetocardiography (MCG) can measure it 20 or more feet away, and its strength is staggering: about 100 times stronger electrically and up to 5,000 times stronger magnetically than the brain's field. Unlike the brain's relatively localised field, the heart's field is expansive, coherent, and influences not only your own physiology but also the people and environments around you.

Vital Outputs: The Power You Become

According to researchers such as Dr. Joe Dispenza, the heart's electromagnetic field is not merely an energy signature; it functions like an interface with the quantum field, the infinite sea of energy and information from which all realities emerge. Your emotions, thoughts, and level of coherence literally shape this field, tuning the signal you broadcast into the quantum realm. This determines how you interact with others, how environments "feel" to you, and even what potential futures are drawn into your experience.

Coherence & Creation

When the brain and heart are in coherence, when your thoughts (electrical) and emotions (magnetic) are aligned, you create a powerful state of resonance. This coherence doesn't just regulate your biology, improving nervous system balance and cellular function; it also amplifies the signal of your biofield.

In spiritual traditions, this state has long been associated with prayer, meditation, and intentional creation. Modern science is beginning to confirm what ancient wisdom has always suggested: coherence allows you to access deeper states of awareness, creativity, and connection with the greater whole. In this state, the heart acts as the bridge between your inner world and the quantum field, where intention and emotion combine to influence not just your personal biology but the reality you experience.

The Potential Role of Melanin in the Biofield

Although there is currently no solid scientific evidence directly supporting this idea, melanin's unique ability to absorb a broad spectrum of electromagnetic radiation, from radio frequencies through visible light, and up to X-rays and gamma rays, opens an intriguing possibility.

Since melanin is also present in the skin's cells and fat layers, it may be capable of absorbing photons generated within the body, such as those UPEs emitted by mitochondria, as they reach the skin's surface. In doing so, melanin could potentially act as a filter or enhancer, reducing incoherent emissions, maintaining coherence within the biofield, and perhaps even re-emitting photons at specific wavelengths, in a manner similar to how internal melanin structures influence mitochondrial biophotons.

The possibility is fascinating, and it would be exciting to see future research explore this hypothesis in depth.

Power Output Three: Structured Water

Inside your mitochondria, the Electron Transport Chain (ETC) does more than just produce ATP; it also manufactures structured water.

At the final step of oxidative phosphorylation, oxygen (delivered to your mitochondria via haemoglobin from the air you breathe) accepts electrons and combines with protons (hydrogen ions) to form fresh, pure water right inside your cells.

This isn't ordinary water; it's deuterium-depleted, structured metabolic water, sometimes called crystalline water or exclusion zone water (EZ water), created molecule by molecule in the inner mitochondrial membrane.

Special Deuterium-Depleted Water

This water is special because it contains only light hydrogen, with almost no deuterium, unlike the water we drink or what's found in the ocean and our blood. In fact, the ocean and our blood plasma

VITAL OUTPUTS: THE POWER YOU BECOME

typically contain about 150 parts per million (ppm) of deuterium naturally.

Deuterium is a heavier isotope of hydrogen, often called "heavy hydrogen", with an extra neutron in its nucleus, doubling the atomic mass of light hydrogen. This bigger, heavier hydrogen doesn't fit well into the tiny rotary motor of ATP synthase; it can jam the spin and clog the mitochondrial energy engine, slowing down nanomotors and enzymes.

Here's where mitochondria act as master water refiners. The inner mitochondrial membrane, packed with proton pumps and nanoscopic channels like ATP synthase and cytochrome complexes, naturally excludes deuterium because its heavier mass moves more slowly. Over time, this selective filtering produces water that is significantly depleted in deuterium.

The Roles of Structured Water

Structured water is the medium of life inside you. Dr. Jack Kruse calls it a chameleon because it seamlessly shifts roles depending on the biological need, conducting electricity one moment, storing energy the next, or shaping proteins into precise molecular machines. It's the most versatile and intelligent water you'll ever "meet," and it exists in its purest form right inside your mitochondria.

Let's explore five of its key roles:

1. Cellular Hydration

Every cell in your body depends on properly hydrated proteins to function. Proteins are dynamic structures, folding and unfolding into specific shapes to carry out their jobs, whether that's catalysing reactions, transporting molecules, or transmitting signals.

- Hydration Layer – deuterium-depleted water (DDW) forms a structured hydration shell around proteins, giving them the flexibility to fold correctly and the stability to hold their shape.
- Electrical Conductivity – Properly hydrated proteins act like biological wires, conducting protons and electrons efficiently. DDW's lighter hydrogen atoms make proton transfer faster and more efficiently, enhancing the speed of your biochemical reactions.
- Molecular Precision – If proteins are under-hydrated or surrounded by deuterium-rich water, folding errors happen. That's when enzymes misfire, signals get scrambled, and cellular energy tanks.

Without DDW, your proteins lose their optimal geometry, like an origami figure collapsing when the paper gets soggy or brittle. With it, they stay perfectly tuned to their tasks.

2. Biological Battery

Structured water has the remarkable ability to capture and store energy from light, especially infrared wavelengths, much like a solar panel converts sunlight into usable power. This energy becomes embedded within the ordered molecular structure of the water and can be released to help drive cellular functions. In this way, your cells maintain an energy reserve that supplements the ATP produced by mitochondria.

This storage capacity comes from structured water's unique properties. It holds a negative electrical charge and organises itself into highly ordered, crystalline-like layers, often hexagonal, along hydrophilic (water-loving) surfaces such as mitochondrial membranes. This ordered arrangement acts as a biological battery, enhancing the cell's ability to store, hold, and deliver energy when needed.

VITAL OUTPUTS: THE POWER YOU BECOME

3. Communication Medium

Structured water does more than store energy; it's also a high-speed communication network inside your body. You can think of it as your body's own "biological fibre-optic gel," guiding light-based and electromagnetic signals between cells with exceptional speed and clarity.

Just as fibre-optic cables transmit internet data as pulses of light with minimal loss, structured water enables the efficient transfer of biophotons and subtle electromagnetic signals with minimal interference or "noise." Its highly ordered, crystalline-like arrangement creates an environment where information can move quickly, reliably, and with precision.

This structure also shapes how molecules find and interact with one another, ensuring that biochemical reactions occur in the right place, at the right time, and with the right partners. In this way, structured water is both the message, the messenger's pathway, and the meeting place for essential cellular conversations, helping your body maintain coherence, synchronisation, and harmony across trillions of cells.

4. Life's Internal Ocean

Inside every cell, structured water forms an "internal ocean" that supports and connects all the body's life processes, just as Earth's oceans cradle and connect life across the planet.

In this protected environment, enzymes work at peak efficiency, speeding up the chemical reactions that sustain life. Proteins fold into their correct shapes, ensuring they function properly, while nutrients and signals travel freely to where they are needed most.

The integrity of this structured water directly influences everything from mitochondrial energy production to DNA repair, and even the stabilisation of the immense voltage across the mitochondrial inner membrane. When this internal ocean is healthy and well-structured, your biology operates with the smooth, balanced rhythm of a thriving marine ecosystem, keeping you resilient, energised, and in sync with life.

5. Buffer against Environmental Stress

Just as Earth's oceans once provided a sheltered cradle for the first life to emerge, structured DDW acts as an internal shield for your cells today. Its ordered structure makes it more resilient against oxidative damage caused by reactive oxygen species (ROS) and more resistant to disruptive forces such as radiation. This built-in stability helps protect the delicate machinery inside your cells, preserving their integrity even during periods of inflammation or environmental stress.

In summary, your mitochondria aren't just energy producers; they're precision distilleries, crafting pure, structured, and energetically charged water that fortifies your cells while powering life's internal currents.

Power Output Four: Carbon Dioxide

Most people think of Carbon Dioxide (CO_2) as just a "waste gas" you breathe out, but biologically, it's far more important. Inside your mitochondria, CO_2 is produced as a byproduct during the Krebs cycle, a key step in breaking down nutrients to harvest electrons and protons for energy production.

Vital Outputs: The Power You Become

Biological Roles of CO_2

- Blood pH regulation: CO_2 dissolves in the blood to form carbonic acid, which helps maintain the body's delicate acid–alkaline balance.
- Oxygen delivery: CO_2 plays a role in the Bohr effect, where higher CO_2 levels help haemoglobin release oxygen where your tissues need it most.
- Vasodilation: CO_2 helps widen blood vessels, improving circulation.
- Neural signalling: Subtle changes in CO_2 levels can influence nerve activity and breathing rhythm.

Circle of Life Connection

The CO_2 your mitochondria produce is part of Earth's grand circle of life. When you exhale it, plants absorb that CO_2 during photosynthesis, using it to build sugars that fuel their growth. Those sugars then become food for animals and humans. In return, plants release oxygen, the very gas we need to power our own cellular respiration. It's a continuous exchange, with every breath linking you directly back to nature.

Power Output Five: Infrared Heat

Your mitochondria also produce infrared light as heat, a byproduct that's actually a vital feature. You can't see it, but you feel it as warmth. Biologically, it:

- Charges your cellular water: Infrared light penetrates deep into tissues and is absorbed by the structured water around proteins and membranes, boosting its negative charge and

energy storage, and constantly energising and structuring it, like recharging a biological battery.
- Keeping your chemistry on track: Most enzymes, the little machines that run your metabolism, only work well in a narrow temperature range. Mitochondrial heat keeps your biochemical machinery in its sweet spot.

Migration & Adaptation

The ability of mitochondria to produce internal infrared heat was crucial for human survival as we migrated from warm equatorial regions into colder climates. Without this adaptation, we could not have maintained stable core body temperature or kept cellular processes functioning efficiently in the cold.

But this adaptation came with a cost. By diverting some of their energy potential into heat, mitochondria sacrificed efficiency in ATP production. In other words, part of the "fuel" that could have gone into powering the body's engines was rerouted to keep us warm. This reflects one of nature's core principles of dissipative structures: the farther life moves from its place of origin, the more it must adapt by dissipating energy differently. For humans, that origin was equatorial Africa roughly 200,000 years ago.

As humans spread farther from the equator, they encountered not just colder temperatures, but changes in light intensity, electromagnetic fields, and the deuterium content of water. Generally speaking, higher latitudes receive weaker, less intense sunlight but also tend to have lower deuterium levels in natural water—nature's way of balancing the biological equation. Electromagnetism, too, varies with geography: deserts often exhibit the weakest geomagnetic strength, while volcanic regions produce the strongest.

Vital Outputs: The Power You Become

This is why latitude exerts such a profound influence on human health and vitality. For example, when someone born in equatorial Africa with dark skin, which is finely tuned to strong sunlight, moves to high latitudes with weak sun exposure, the environmental inputs their mitochondria evolved to expect are suddenly absent. This mismatch can disrupt cellular signalling, metabolic pathways, and overall function. One striking study noted that Somali immigrants in Detroit, far from their equatorial homeland, experienced one of the highest autism rates recorded, something virtually unheard of in Somalia itself.

The lesson is clear: your environment shapes how your mitochondria work, sometimes dramatically, sometimes subtly. Every shift in light, magnetism, temperature, or water composition reshapes your internal biology and influences how well you can thrive in a given place.

The Perfect Input & Output Loop

When you live in sync with the fundamental forces of nature, your mitochondria awaken to their fullest potential. This is the optimal Input–Output Loop, nourishing every part of you, inside out. For a summary of these vital inputs and outputs, see Appendix A.

Every ray of nourishing sunlight, every sip of living water, every breath of oxygen-rich air, and every moment in a clean electromagnetic environment delivers flawless signals to your mitochondria. In return, they gift you boundless energy, pristine structured water, coherent biophotons, and electromagnetic fields within and around you. These outputs naturally feed back into the inputs, creating a self-sustaining loop of vitality. This is the harmony that fuels an optimal life.

See yourself as a giant atom, your body as the nucleus, your electromagnetic field as the electron cloud. Your feedback loops hum in perfect rhythm, constantly sensing, adjusting, and aligning you with the living environment around you.

This is the foundation of the Atomic Glow. When you master this loop, you're no longer just surviving, you're radiating. You are a living testament to the truth: you are nature, illuminated from within.

CHAPTER 12

MODERN DISRUPTORS OF YOUR GLOW

*"The further you get away from nature,
the more lost you become."*
Dr. Jack Kruse

Modern life, especially over the past few decades, has steadily pulled us further away from the natural glow we were designed to radiate: mental clarity, resilience, and that deep, vibrant feeling of being in love with life.

Alongside the rising rates of cardiovascular disease, neurodegeneration, and autoimmune disorders that are rooted in the body's most energy-hungry organs (the heart, brain, and gut), our overall health span has steadily declined in recent decades. Since 2010, rates of anxiety and depression have surged by more than 25% worldwide. The global prevalence of type 2 diabetes has more than doubled in the last 30 years, now affecting over 10% of adults in many countries. Brain fog has become so common that one in three adults struggles with daily concentration.

Despite access to countless health products, supplements, and wellness advice, many feel stuck, frustrated that they aren't thriving even though they're "doing everything right."

If you've followed the previous chapters on the Input–Output Loop, you already know where to look: it all begins with the quality of the inputs we feed our biological engines.

So let's start by shining a light on the very first disruptor, light itself.

LIGHT CATASTROPHE

The Sunlight-Deprived Modern Life

Sunlight is the spark that ignited life on Earth and the essential force that keeps the integrity of our biology. Yet, in modern life, we spend most of our time shutting out the Sun, from almost sunrise to sundown.

The Indoor Trap

Studies show that the average person now spends up to 95% of their time indoors, largely disconnected from natural sunlight. Even when sunlight filters through windows, modern glass blocks most of the ultraviolet (UV) and all of the infrared (IR) light, two critical wavelengths that power our cellular biology and keep our internal clocks operating precisely.

Modern Disruptors of Your Glow

Additionally, indoor spaces are often associated with weak lighting and poor air circulation, creating dampness, the perfect conditions for mould and harmful microbes to thrive. This invisible threat also contributes to chronic inflammation, respiratory problems, and a decline in overall well-being.

Sun-Deprived Eyes & Skin

When we step outside, we often block the very signals our bodies crave from entering the eyes and skin, our two primary photoreceptive organs. Sunglasses, while protecting us from glare in extreme conditions, shield our eyes from the natural light cues essential for synchronising circadian rhythms, the internal clocks that regulate sleep, hormone cycles, and mood. Without these signals, our biological timing becomes disoriented.

Modern vision-correcting technologies like LASIK surgery can further disrupt the eye's natural ability to process light. Disturbingly, some studies link LASIK with the sudden onset of depression and suicidal thoughts, likely tied to altered light-based biological signalling. Contact lenses, meanwhile, not only block sunlight but also reduce oxygen flow to the cornea, impacting eye health and function.

Our skin, the largest organ and a vital sunlight receptor, is often hidden beneath clothing and coated with synthetic sunscreens that block the life-giving wavelengths our cells need to thrive.

Sun-Deprived Food

Even the food we consume has lost its natural connection to sunlight. Processed under artificial lighting, these foods lack the rich electron energy that sunlight-grown foods naturally carry. This imbalance—more protons, fewer electrons—disrupts cellular energy production, slowing metabolic function and breaking down the delicate harmony within our bodies.

All these layers of sunlight deprivation, indoors and outdoors, in our environment, and on our plates, combine to block the vital light that fuels and harmonises every cell in your body.

The Blue Light–Illuminated Modern World

We're not just starving ourselves of the Sun's nourishing light; we've also flooded our lives with artificial blue light, which creates another devastating layer of disruption that confuses and derails our biology.

What Is Blue Light?

Blue light is a high-energy, short-wavelength light naturally found in the visible spectrum of sunlight that plays a crucial role in human biology. It signals your brain that it's daytime, helps regulate your circadian rhythm, and prepares your body to be alert, active, and ready to face the day.

But here's what many don't realise: blue light itself is a biological stressor. Nature understands this perfectly. That's why the Sun never shines blue light alone, but always pairs it with calming wavelengths like infrared, red, and UVA light. These gentle companions balance blue light's intensity, preventing your body from slipping into a constant "fight or flight" state.

What once lit our days gently from the sky now bombards us in isolation round the clock, from our homes and workplaces to the streets outside. Indoor lighting, along with modern devices and appliances like smartphones, tablets, TVs, cars, and even refrigerators, all emit this intense, isolated artificial blue light.

Modern Disruptors of Your Glow

How It All Started

It began in 1893 with the invention of the electric light bulb. The first incandescent bulbs glowed warmly, mimicking the spectrum of natural sunlight, with calming red and infrared wavelengths that supported our biology.

As technology advanced, the light in our daily lives changed dramatically. Modern LEDs, fluorescent bulbs, and digital screens pump out intense blue light while stripping away the natural red and infrared balance. This shift, marketed as "energy saving", comes with no solid scientific backing but plenty of biological consequences.

How Blue Light Hijacks Your Biology

Your body, finely tuned for the rhythms of sunlight and darkness, was never designed to endure this constant biological stress.

1. It Damages Mitochondria

When you're exposed to artificial blue light, without the presence of balancing red, infrared, or UVA light, your mitochondria face overwhelming stress. Here's what happens inside your cells:

Blue light increases the production of reactive oxygen species (ROS) within mitochondria. These unstable molecules damage mitochondrial membranes, proteins, and DNA when they accumulate excessively.

Under natural sunlight, infrared and red wavelengths activate cytochrome c oxidase, a vital mitochondrial protein responsible for producing cellular water and supporting efficient energy generation. Artificial blue light, however, suppresses this process and overwhelms your mitochondria with damaging stress.

Chronic exposure to artificial blue light causes mitochondrial dysfunction. This means your cells can't produce enough ATP energy, structured cellular water, or coherent biophotons and electromagnetic fields necessary for optimal function.

This leads to:

- Low energy: You feel fatigued, weak, and mentally foggy.
- Increased inflammation: Excess ROS damage tissues, triggering immune responses that worsen over time.
- Accelerated ageing: Damaged mitochondria struggle to repair DNA, increasing the heteroplasmy rate and speeding up cellular ageing and disease.
- Disrupted cellular communication: Chaotic biophoton emissions lead to poor cellular signalling and an incoherent bioelectromagnetic field, throwing off your body's internal harmony.

When your mitochondria fail, it's not just low energy you experience; it's a collapse across your entire biological system. Hormones go out of balance, your immune defences weaken, brain function declines, and your heart struggles to maintain the vital rhythm that sustains life.

2. It Degrades Melanin

Compromised Energy Producing Systems

Emerging research suggests that melanin, the pigment that gives skin its colour, can function like a natural solar panel, capturing sunlight and converting it into usable energy through processes that may even bypass the mitochondria. This light-harvesting ability can act as a backup energy source, complementing the energy we generate from food.

Modern Disruptors of Your Glow

Excessive artificial blue light, however, can degrade melanin, weakening this light-based support system and forcing your mitochondria to work harder to meet your body's energy demands. As discussed in the previous section, artificial blue light also directly damages mitochondria, creating a double blow to your body's energy production.

Disrupted Cellular Communication

Internal melanin acts as a master communicator and conductor within your body. Under healthy conditions, it absorbs virtually every frequency of light emitted by your cells, then captures, interprets, and redistributes that light. This precise coordination keeps your biological systems in harmony.

When melanin is bleached or degraded by excessive artificial blue light, this communication network is disrupted. As a result, cellular synchronisation weakens, and your biology could shift from coherence to chaos. In addition, damaged melanin can release vitamin A derivatives, which, in excess, generate localised oxidative stress. This stress can harm surrounding cells, setting off a cascade of damage and dysfunction throughout tissues.

Biological Consequences

This spiralled, disrupted signalling affects everything from immune function to hormone regulation, even influencing higher consciousness.

- Immune System: Melanin plays a critical role in light-mediated immune signalling and cellular defence. When damaged, your immune response slows down or becomes erratic, leaving you vulnerable to infections and chronic inflammation.
- Skin Health: Melanin protects skin cells by neutralising reactive oxygen species. Blue light exposure without adequate

protection accelerates premature ageing, hyperpigmentation, collagen breakdown, and photoaging.
- Neurological Health: Melanin is concentrated in brain regions like the substantia nigra, which is vital for dopamine production and motor control. Its degradation is linked to neurodegenerative diseases such as Parkinson's and other dopamine-related disorders.
- Heart Health: Melanin is present in the heart and blood vessels, where it contributes to protecting cardiovascular tissues from oxidative damage and supports nitric oxide signalling critical for vascular tone and blood flow. Melanin degradation can impair these protective effects, increasing the risk of heart disease and poor circulation.

3. It Disrupts Oxygen Delivery

Artificial blue light doesn't just damage your energy systems and cellular signalling, it also sabotages your body's ability to use oxygen effectively.

As we discussed earlier, haemoglobin, the molecule in your blood responsible for carrying and delivering oxygen, depends on iron at its core in the right form to do its job. This iron needs to be in the Fe^{2+} (ferrous) state to bind and transport oxygen efficiently.

But artificial blue light causes oxidation, flipping that iron from Fe^{2+} to Fe^{3+} (ferric), a form that can no longer hold onto oxygen. So even if you're breathing perfectly, your blood loses its capacity to deliver oxygen where it's critically needed.

Oxygen is essential, and your cells need it like a fire needs air. Each breath you take is meant to fuel your mitochondria. Without oxygen, the electron transport chain within these tiny powerhouses grinds to a halt. When this critical process stalls, your cells can no longer produce the life-giving outputs that sustain you: cellular water, energy, coherent biophotons, and your vital electromagnetic field.

Without this essential flow, thriving becomes almost impossible. Instead, your body is left struggling just to survive.

4. It Makes You Addicted

From a psychological perspective, artificial blue light can become addictive, trapping you in a cycle, like how casinos in Las Vegas use light to hook gamblers. This isn't just about habit; it's about how your brain chemistry gets hijacked.

In natural sunlight, a molecule called POMC gets activated and triggers the production of beta-endorphin, a natural opioid that provides a healthy sense of reward, calm, and well-being. It also helps regulate dopamine, a neurotransmitter that governs motivation, pleasure, and mood balance.

When you're deprived of sunlight, your body produces less beta-endorphin. This leaves a gap, a craving for pleasure and stimulation that your brain desperately seeks to fill. That's when artificial blue light from screens and devices becomes a tempting substitute, offering quick dopamine hits but without the true balance and healing that natural light provides.

This creates a vicious loop: the less sun you get, the more your brain craves artificial stimulation, making it harder to break free and return to a lifestyle that truly supports your health and vitality.

EMF CATASTROPHE

Lack of Natural EMF

We are not only photonic in essence, we are also electromagnetic beings. Remember that tiny atom from Chapter 7? The orbiting electron cloud, governed by the electromagnetic force, shapes the electrons' behaviour. Together, they determine the atom's size, shape, and chemistry. Since we are made of atoms, the electromagnetic field we are exposed to also shapes our relationship with sunlight, and together, they influence everything that happens within our bodies.

As living creatures of the Earth, humans are designed to stay in sync with the Earth's electromagnetic field. For millions of years, humans lived in constant contact with the Earth's surface, absorbing its natural frequencies, most notably the Schumann resonance, a gentle 7.83 Hz signal that keeps our brainwaves, heart rhythms, and cellular communication in harmony. We also relied on the Earth's geomagnetic field, which our bodies use to orient themselves, regulate blood flow, and maintain internal balance.

Today, that connection is broken. Modern life has pulled us out of the electromagnetic harmony we evolved in. Rubber-soled shoes insulate us from the Earth's conductive surface, blocking the flow of these beneficial frequencies. Indoor living, on wooden or carpeted floors, and in high-rise buildings, further removes us from direct ground contact. As a result, our cells operate without the steady electromagnetic cues they once depended on.

Chronic Exposure to Non-native Electromagnetic Fields

At the same time, we've blanketed the planet and ourselves with non-native electromagnetic fields (nnEMFs), from Wi-Fi and cell towers to power lines running above and beneath the ground. These artificial signals disrupt the Earth's natural frequencies and replace them with patterns our biology was never designed to navigate.

What is nnEMF?

The electromagnetic spectrum spans a vast range from extremely low frequencies at one end, through radio frequencies (RF), microwaves, infrared, visible light, and ultraviolet, all the way up to X-rays and gamma rays at the high-energy end.

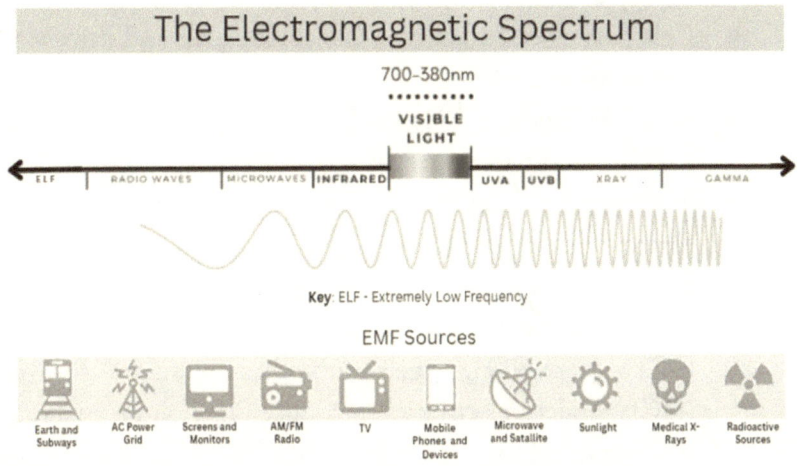

Graph 3: The Electromagnetic Spectrum

Yet only a small slice of these frequencies naturally reaches Earth's surface. Thanks to the planet's magnetosphere and atmosphere, most harmful radiation is deflected or absorbed. What gets through are the frequencies that life evolved under and depends on, primarily sunlight in the visible spectrum, along with beneficial ultraviolet and infrared wavelengths.

NnEMFs are man-made electromagnetic signals that fall outside this natural, life-supporting range, occurring at unnatural intensities and patterns. For nearly all of human history, these artificial frequencies didn't exist. But in the last 150 years, especially the past three decades, we have flooded our environment with EMFs spanning almost the entire spectrum.

Here's a breakdown of the main sources and where they sit on the spectrum:

- Cell Phones & Mobile Networks – RF and microwaves, from hundreds of MHz to tens of GHz (4G to 5G).
- Cell Towers & Satellites – Continuous RF and microwave emissions.
- Wi-Fi, Bluetooth & Wireless Devices – Microwaves, typically 2.4 GHz and 5 GHz bands.
- Microwave Ovens, Smart Meters & Smart Appliances – Microwaves, often around 2.45 GHz.
- Power Lines, Home Wiring & Dirty Electricity – Extremely low frequencies power grid running on alternating current (AC) at 50 or 60 Hz, where our bodies run on direct current (DC). Modern electronics introduce rapid on-off switching that creates bursts of high-frequency "dirty electricity" in the kilohertz to megahertz range, adding chaotic electrical noise to our home environment.

Biological Impact of nnEMF

Electromagnetic fields influence every moving charged particle. Since living systems are full of conductive fluids and charged ions, EMFs interact with virtually every process in biology. When we surround ourselves with constant nnEMFs, our biology starts to experience both subtle and obvious disruptions.

Modern Disruptors of Your Glow

1. It Damages Your Mitochondria

Mitochondria are exquisitely tuned electrical systems. They need a precise voltage gradient, an orderly flow of electrons and protons, and a low-entropy (low chaos) environment to function at their peak. Experts like Dr. Paul Héroux, a toxicology professor at McGill University, have highlighted that even low-level exposures, such as those common from power-frequency fields, can impair mitochondrial function, increase oxidative stress, and affect cellular metabolism.

Non-native EMFs impact your biological systems at their core. They:

- Short-circuit the voltage gradient, the tiny electrical charge across the mitochondrial membrane that powers ATP production. Without that spark, the whole energy assembly line grinds to a halt.
- Jam the cell's communication lines by overstimulating voltage-gated calcium channels. Suddenly, calcium pours into the cell in excess, like a flood overwhelming a city, disrupting normal signals and setting off damaging chain reactions.
- Amplify oxidative stress, causing reactive oxygen species to leak and damage your proteins, fats, and DNA.
- Reduce ATP energy output. With the membrane potential impaired, ATP production drops sharply. Your cells start running on an empty tank.
- Dehydrate cells from within. Mitochondria also make structured "exclusion zone" water. When this drops, cells lose hydration, detox slows, and nutrient transport and communication suffer.

When mitochondria are compromised, every system in your body struggles. At the fundamental level of life, energy is life's currency. Without it, repair slows, resilience weakens, and healing becomes an uphill battle.

2. It Distorts the Biofield & Dims the Inner Light

As we explored earlier, every cell in your body emits both biophotons and generates an electromagnetic field through mitochondria. This light and energy are fundamental to cell-to-cell communication, coherence, and even consciousness itself.

NnEMFs disrupt this delicate system. They act like static in your body's electrical wiring, distorting your natural frequencies, including brainwaves and heart rhythms. As a result, your inner system falls out of tune. The nervous system struggles to regulate, hormone signals become inconsistent, and the light-based communication that fuels higher thought, emotional stability, and creative insight begins to fade.

When your biofield is disrupted, it's harder to feel aligned with yourself and the world around you. Practices like meditation or prayer can feel disconnected, not because you've lost your touch, but because the internal signal is filled with noise. Your connection to the quantum field, your intuition, and your sense of the Divine is interrupted.

3. It Disrupts Oxygen Delivery

Haemoglobin, the molecule in red blood cells responsible for carrying oxygen, is highly sensitive to electromagnetic fields. This is because oxygen itself is paramagnetic, meaning it's naturally attracted to EMF fields.

When you're in a high nnEMF environment, this attraction pulls and redistributes oxygen unnaturally, throwing off the body's precise delivery system. As a result, your tissues may not receive oxygen where and when they need it most. This disruption can lead to chronic fatigue, slower recovery, and even anxiety, not due to a true oxygen shortage, but because your cells can't access it efficiently.

Modern Disruptors of Your Glow

4. It Disrupts the Relationship with Natural Light

NnEMFs change the way our body absorbs and interprets sunlight. For example, you might live in a sunny place like Los Angeles, soak up hours of sunshine, and even develop a golden tan. But when you test your Vitamin D levels, they still come back low.

That's because your skin's ability to convert UVB rays into vitamin D depends on a stable electrical environment. When you're surrounded by nnEMF from cell towers, Wi-Fi, Bluetooth, and other sources, your voltage-gated calcium channels (VGCCs), the tiny gatekeepers that control the flow of calcium into cells, become overstimulated.

Calcium isn't just for bones and teeth; it's the universal signal switch for the brain, heart, muscles, mitochondria, and hormones. When VGCCs are forced open too often, calcium floods in, sparking oxidative stress, inflammation, and mitochondrial damage. This disruption alters your body's ability to process sunlight properly. You might see the Sun, but your cells can't.

CHEMICAL CATASTROPHE

Modern life has brought convenience, but often at the cost of our body's most fundamental energy systems. One of the biggest disruptors is man-made chemicals. These are elements and compounds that biology never selected, substances that our cells have no evolutionary blueprint to handle. Examples include heavy metals like mercury and lead, fluoride, aluminium, endocrine-disrupting plastics like BPA, and persistent pollutants like PFAS ("forever chemicals").

They're everywhere and often impossible to avoid completely:

- In our food and water. From pesticide and herbicide residues on produce, to chlorine, fluoride, and even microplastics in the water supply, plus artificial additives and preservatives in packaged foods.
- In products we use every day, such as sunscreen containing chemical UV filters, makeup with synthetic pigments and parabens, and household cleaners loaded with volatile organic compounds (VOCs) that linger in the air we breathe.
- In health-related products, including pharmaceuticals, some vaccines and synthetic supplements that don't match the forms naturally found in whole foods.

Once inside the body, many of these toxins interfere directly with the mitochondrial functions, while generating oxidative stress that accelerates cellular damage. They can accumulate in tissues and interfere with the very inputs and outputs of mitochondria. For example, reducing oxygen delivery, disrupting nutrient pathways, generating excess free radicals, and impairing ATP production. This also disrupts hormone balance and cellular communication, creating a persistent, low-grade stress on our biology that quietly erodes health over time.

The Sea of Constant Stress

These disrupters create a constant undercurrent of stress inside your body. Unlike short bursts of healthy stress that make you stronger, these disruptors act like background static, always on, always draining.

They confuse your circadian timing, disrupt your cellular communication, and burden your detox systems. Over time, this low-grade stress doesn't

Modern Disruptors of Your Glow

just tire you out; it interrupts the entire ecosystem inside, from energy production to expression of DNA.

Your glow is dimmed not because you lack anything, but because your environment is working against you. The good news? By recognising these disruptors for what they are, you can implement appropriate strategies to reclaim the power to quiet the static, reset your biology, and let your natural radiance shine again.

CHAPTER 13

NATURE-INSPIRED MODERN LIVING

"Vitality and beauty are gifts of nature for those who live according to its laws."
Leonardo Da Vinci

Living in an unnatural mix of missing essentials and overwhelming disruptors, too little natural light, too much artificial light, constant non-native EMFs, and a daily cocktail of man-made chemicals is biologically expensive.

Back to Basics

If we zoom in to our most fundamental level, the atom, we see the truth. You are, at your core, a nucleus surrounded by an orbiting field where electrons dance with light. In nature, that photonic and electromagnetic field is harmonious, tuned to the frequencies that shaped life itself.

But today, that field has been crowded with foreign signals and chaotic patterns, an orbit of stress that reshapes what's happening inside your body. And just as the electron field influences the nucleus, this stressed outer environment distorts your inner environment, making it equally stressed and chaotic.

To renovate them, we have to restore their inputs. That means feeding them what they evolved to thrive on: the rhythms, light, and frequencies of the natural world, while minimising the modern disruptors that weigh them down.

What follows is your roadmap to living a nature-inspired modern life. It's about immersing yourself in the healing forces of nature, shielding your body from modern disruptors, and rebuilding a thriving colony of mitochondria, so you can once again run at full power, with clarity, vitality, and unshakable resilience.

LIGHT RENOVATION

Embrace the Sun

Coming from a culture where fair skin is considered the ultimate symbol of beauty, tanned skin was looked down upon, seen as a mark of the working class, the farmers. So when I first moved to New Zealand, I was that girl hiking through the woods with an umbrella in hand, terrified of letting the Sun touch my skin. Learning to step into the Sun and embrace a tan was a huge challenge for me. But as Maya Angelou said, "When you know better, you do better." Eventually, nature's wisdom replaced inherited beliefs.

NATURE-INSPIRED MODERN LIVING

As I deepened my understanding of light and the critical role our eyes play in human biology, another truth revealed itself: the hidden cause of my severe postnatal depression. The answer traced back to a choice I made, getting LASIK surgery after years of wearing contacts. It was one of my greatest wake-up calls. You should never interfere with your eyes, because when you alter your eyes, you also alter your brain. And when you interfere with your brain, you disrupt one of your most vital tuning forks as a human being. We'll explore this profound connection in the next chapter.

I haven't missed a sunrise since the day I first heard Dr. Jack Kruse speak about the magic of morning light, and I never will for the rest of my life. That first light of day isn't just beautiful, it's a biological reset button. It's the secret switch for mitobiogenesis, the process where your body creates brand new mitochondria and prunes away the weak ones. Morning sunlight is nature's quality control, making sure only your healthiest energy engines stay in the fleet.

When I step outside, skin bare and eyes open to the miraculous rising sun, I'm not just seeing and feeling the magic, I'm feeding my body the most fundamental nutrient there is, pure light. Of course, I ditched sunscreen and sunglasses that block these life-giving nutrients from entering my eyes and skin.

To make connecting with nature a sustainable way of life, we made a bold, heart-led investment: we built our home near the beach, surrounded by lush forests, rolling green hills, and rich wildlife, so the healing touch of the Sun and nature are never far from reach.

Every morning at dawn, I slide open the wide glass doors and welcome the first magical beams of red-golden light as they pour over my skin and into my eyes. I'm instantly bathed in the grandest red-light therapy ever created, one designed by the universe itself. After rising, I take my daily walk on the beach, letting my photoreceptors continue to drink in the nourishing sunlight as the day awakens around me.

Outside, we've cultivated abundant organic food forests, where fresh, seasonal fruits and vegetables grow in harmony with the land and under the power of the Sun. These foods are more than nourishment; they are living packets of electrons and protons, directly charging our mitochondria with the very same light that nurtured them, creating coherence and harmony within us.

To me, gardening is far beyond growing food. It's a perfect outdoor lifestyle, immersed in the sunlight, surrounded by living plants that give abundant oxygen and reflect rich infrared light, the light that fuels mitochondria, back to us. It's the way of connecting with Mother Nature, observing and learning nature's timeless wisdom.

If life doesn't allow for gardening, a beach walk, or a house designed around sunlight, you can still embrace the Sun in powerful ways. Nature is generous and meets you wherever you are. Step outside first thing in the morning, even for just a few minutes, and let the rising light touch your skin and enter your eyes. Take your coffee or tea by an open window, or walk around the block without sunglasses. Open your curtains and windows wide during the day, or better yet, work near a window so natural light fills your space. If you live in an apartment or a busy city, lunch breaks in the park or short walks outdoors can still tune your circadian rhythm and charge your mitochondria. It's not about perfection, but consistency. The Sun is always there; it only takes you to show up and receive it.

Nature-Inspired Modern Living

Minimizing Artificial Light Exposure

The best light during the day is sunlight. The best light at night is darkness. Any artificial light after sunset is foreign to our biology and disrupts circadian timing to some degree. Following nature's natural light–dark cycle will always be the gold standard. When that isn't possible, the next best option is to use lighting that balances or filters out excess blue wavelengths. Still, it's important to remember that all man-made lighting, even popular tools like red-light therapy or photobiomodulation, relies on isolated frequencies that only mimic a narrow slice of the Sun's spectrum. Use them mindfully.

For all light fittings in our home, we chose warm amber or red light bulbs that mimic sunrise and sunset. They are also completely flicker-free to reduce the pulse that causes biological stress as in normal lighting. In long summer days, we never needed to use them. Before moving into this home, we relied on table lamps with the same features, or simply lit candles when light was required, keeping our evenings soft and natural.

We installed large windows and sliding glass doors that can be opened all year round to flood the house with daylight, so even indoors, we're bathed in natural light.

Screens such as laptops, tablets, and phones are always dimmed to amber tones. They also share space with open windows or doors to let sunlight balance their artificial glow.

Even the harsh white fridge light is wrapped in ruby film, and when driving, we crack open the car windows to let in the real light.

When the Sun goes down, we follow its lead. Minimal screens, minimal artificial light. Nighttime darkness is just as vital as daytime light.

Our cells sense the absence of the Sun; that's their signal to shift into relax-and-repair mode. By embracing the dark, we allow our engines to carry out their nighttime work as nature designed.

In the short days of winter, when work or connection calls for a brief burst of evening screen time, we slip on our blue-light-blocking glasses and switch on soft red light to balance the impact of artificial blue light. Limited reading, especially for our two little bookworms, happens by the fire, under red lamps and with blue-light-free amber bulbs.

In essence, our lives are fueled by natural light, even indoors. But this transformation didn't happen overnight. We built these rhythms one small habit at a time, until they became part of our daily lives. Today, they feel effortless, automatic, natural, and deeply life-giving.

EMF RENOVATION

Grounding or Earthing

Grounding to Earth's natural electromagnetic field is one of the most powerful daily practices in our lives. I spend most of each day connected to the Earth.

Barefoot at the beach, gardening, or playing with my children on the lawn, my body is constantly absorbing free electrons from the Earth and syncing with her gentle Schumann resonance, the very heartbeat of our planet.

Even our home was designed with grounding in mind. We chose polished concrete flooring in the main living areas, so we stay grounded indoors as well.

Nature-Inspired Modern Living

One important note: not all ground is created equal. In areas where the power grid is buried underground or cell towers are nearby, grounding may not be ideal. Always choose natural surfaces such as sand, soil, or grass, away from man-made electrical interference for the purest connection.

Minimize nnEMFs

We've designed our home and lifestyle to minimize exposure to non-native EMFs as much as possible. Our house was built with a low-EMF principle in mind: the internet is fully hard-wired with Ethernet cables, so Wi-Fi is never needed. Even my phone connects through Ethernet when I go online. In previous houses, we turned off Wi-Fi and ran long Ethernet cables across rooms for internet connection.

I rarely use mobile data, so it stays switched off, removing unnecessary frequencies from nearby cell towers. When I listen to downloaded podcasts or audiobooks, my phone is always in airplane mode, so it isn't constantly pulsing signals to mobile networks. (One note: some phones still emit radio frequencies even in airplane mode, so it's best to check with an EMF meter to be sure.)

We also paid close attention to how electricity was wired throughout the house. For example, all wiring to each bedroom runs to a central point far from where we sleep, minimizing electrical fields around the bed. In homes where wiring can't be controlled or verified, simply moving your bed farther from the wall can make a meaningful difference. Thanks to the Inverse Square Law, the strength of electromagnetic fields drops dramatically with distance.

We had the communication chips removed from the smart meter, and we only purchase appliances without RFID chips, and with the option to fully disable Bluetooth and Wi-Fi.

ATOMIC GLOW

Optimising Light and EMF Outside the Home Environment

Navigating environments outside the home can sometimes be challenging. Offices, libraries, and most public spaces in the city are often saturated with EMF and artificial blue light. If you spend most of your week in these environments, for example, working in an office five days a week, you may need to take extra steps to protect your biology.

Practical strategies include:

- Adding UVA and red-light table lamps to counterbalance the intense blue light.
- Wearing amber glasses and dimming screens.
- Using wired internet instead of Wi-Fi whenever possible.
- Ensuring all devices are properly grounded.

When you leave these environments, make it a ritual to reconnect with nature, go barefoot to ground yourself, and soak up natural light.

Cold exposure is another powerful way to supercharge your mitochondrial engine. When your body is exposed to cold, mitochondria are stimulated to work more efficiently, producing more heat and energy through a process called thermogenesis. This not only strengthens your metabolic flexibility, your ability to switch between burning fat and glucose, but also encourages the growth of new, healthier mitochondria. Remarkably, cold also signals mitochondria to increase the release of ultra-weak biophotons, vital for cellular communication and repair. In short, cold acts as a gentle stressor that trains your cells to become more resilient, enhancing energy, fat-burning, and overall vitality.

Ultimately, I see it as an 80/20 rule: if we do the right things most of the time and give our best effort, we're far better off than if we didn't.

Nature-Inspired Modern Living

One silver lining from the Covid pandemic is the rise of remote work. Even being able to work from home part of the week helps tremendously, since the home environment remains the most important factor in managing nnEMF, especially at night when the body does its deepest repair and reset during sleep.

Returning Closer to Our "Home"

Some people may still struggle even when they do everything right, watch the sunrise, ground daily, minimize artificial light and EMFs, and eat local food. For those with significantly compromised health, or for those whose natural design is mismatched with their current environment, there may be another option: moving closer to home.

By home, I mean closer to the equatorial regions, where human life first emerged. There, the Sun shines with greater intensity year-round, and the electromagnetic environment is more stable. These inputs are the foundation that shaped human biology over millions of years.

For darker-skinned individuals, this truth is even more important. Their skin was designed for strong, consistent equatorial sunlight. Living at high latitudes with weaker light and harsher seasonal swings places a constant stress on their biology. The further we move from our ancestral environment, the more compensation the body requires, and for some, that compensation comes at the cost of health.

It may not be an option for everyone, but for those whose health has reached a breaking point, moving closer to home can be one of the most powerful healing strategies, restoring harmony between body, environment, and the forces of nature that originally shaped us.

CHEMICAL RENOVATION

Since embracing a more natural lifestyle, I began questioning everything in our lives, the things we used, the habits we followed, even the beliefs we were taught from a very young age.

Whenever I came across something man-made, I asked myself a simple question: "How does nature do it?" That one question began peeling back the layers of the life I had lived for decades.

I still remember the first moment clearly; it was with a tube of nappy cream. A popular brand, one recommended by midwives and pediatricians. But when I turned it over and read the ingredients, I was stunned: pesticides. How could something with pesticides be considered safe for a baby? That was the last day I ever used it. Straight into the bin.

Nappy cream was a new invention because our parents never used it on us, and we turned out just fine. So I asked, how does nature do it? The answer was simple: fresh air. From then on, after every nappy change, I let my babies' skin breathe, air dry. Not once did they develop a rash.

That moment cracked the door open.

The next was sunscreen. When I look at the long list of chemicals, many linked to hormone disruption and cancer, I couldn't justify putting it on my skin, knowing it would be absorbed into my bloodstream, into my cells, into my mitochondria. We've been sold the idea that we need to be "protected" from the very sun we are designed to thrive in. But the truth is, nature already gave us protection: morning sunlight. The red and UVA rays in the morning sun prepare and condition our skin before UVB arrives later in the day. UVB on its own can be harmful, yes. But nature never delivers it alone; it always comes balanced with nourishing red and infrared light. Nature is intelligent, far beyond human comprehension.

Then came the microwave oven. Remember covering a steak with a damp cloth so it doesn't dry out? That's because microwaves strip water from food. Can you imagine what happens when our cells are exposed to microwave radiation? The same thing, cellular dehydration. Knowing the critical role structured water plays in our biology, it was an easy decision: no more microwave oven.

From there, the unraveling picked up speed. Non-stick pans releasing toxins when heated. Harsh chemical cleaning sprays filling our homes with invisible stressors. Skincare and makeup products packed with synthetic ingredients our grandparents never needed.

One by one, they all went.

Now, our home runs simply: cleaning with baking soda, vinegar, lemon, or lavender; cooking with stainless steel, cast iron, glass, and ceramic, using coconut oil or tallow. For skincare, it's nothing more than pure plant oils like rosehip, or sometimes, nothing at all.

Even our water comes from underground concrete rainwater tanks, free from fluoride and chemical treatments. Our garden thrives without synthetic fertilizers, just compost, seaweed, and biodiversity, which together create the natural harmony that supports growth.

And here's the truth I discovered: over 99% of modern products that are loaded with chemicals are unnecessary. Most were invented for convenience, or worse, to solve problems created by other man-made products. It's a vicious loop. But once you begin living by the laws of nature, you realise how little you truly need, because the problems you once faced simply disappear.

Nature doesn't need fixing. We just need to stop interfering.

Healthy Living in the Modern World

Healthy living in the modern world is not only possible, it's your birthright. Despite the noise of technology, the lure of convenience, and the overwhelm of conflicting health advice, your body still speaks the timeless language of nature. It remembers the rhythm of sunlight and darkness, the nourishment of clean water and whole foods, the grounding pull of the Earth, and the invigorating reset of cold.

The key is not to chase perfection, but to live in conscious alignment with these natural laws most of the time. By reclaiming your light environment, minimizing unnecessary EMFs, eating real food grown in harmony with nature, and honouring rest and connection, you restore your biology to what it was designed for: vitality, resilience, and radiant energy.

The modern world doesn't have to work against you. When you learn to master its challenges and return to nature's wisdom, you don't just survive, you thrive. Healthy living is not about restriction, struggle, or fear. It's about remembering who you are, honouring the design of your biology, and stepping fully into the life you were born to live.

CHAPTER 14

THE INNATE TUNING FORKS OF REALITY

"Reality is born of potential, but defined by the frequency you tune."
Alison Davis

Imagine this: you've rebuilt your mitochondria, the tiny engines of life, by aligning with nature's laws, soaking in natural light, moving with the rhythms of day and night, and clearing away the noise of artificial frequencies and chemicals. Your body hums with energy and precision. Every cell is in tune, every rhythm on beat, chaos and order moving like a perfect dance. This is the body nature designed.

For chimpanzees in the wild, this would be all they need to navigate the world. Strength, agility, and survival are their masterpieces. But along the arc of evolution, we traded a larger gut for a larger brain. We gained two remarkable frontal lobes, the quantum processors capable of imagination, creativity, and vision. Our hearts, too, expanded, not just as pumps but as the centres of magnets, with each cell housing thousands of quantum engines that connect us to something greater.

The Free Will Paradox

Hardware alone does not guarantee happiness, and this is where free will enters the picture. Unlike animals, who live in natural alignment by instinct, humans have been given something far more powerful: the ability to choose. We can choose what to think, how to feel, and how to direct the energy within us. With every choice, we paint our reality, sculpting destiny with the tuning forks of thought and emotion.

That's why, even throughout history, when people lived close to nature and free from the modern disruptions we face today, some still suffered while others experienced joy. The difference wasn't only the environment; it was how they tuned.

Free will is like a set of tuning forks. No matter how much energy your body can generate or how powerful your biofield can become, free will has the final say. Each thought, each emotion, is a choice that shifts your frequency. Like music, that frequency can harmonise into beauty, or collapse into dissonance.

The integrity of this tuning depends, first and foremost, on alignment. The more you live in harmony with nature's design and the more your hardware unlocks its highest potential, the clearer, stronger, and more precise your tuning becomes. But when you drift out of sync, surrounded by modern disruptors like artificial light, nnEMFs, toxins, and chronic stress, your inner tuning weakens. Noise replaces clarity.

The Innate Tuning Forks of Reality

Instead of amplifying harmony, you may find yourself resonating with chaos.

Alignment with nature isn't just helpful; it is the foundation. It determines whether your inner frequencies rise into order and coherence, or spiral into disorder and confusion.

The Tuning Mind: Thoughts & Beliefs

Our thoughts and beliefs are one of the most powerful tuning forks of free will. They shape the stories we tell ourselves, the judgments we make, and the way we interpret circumstances and experiences. Whether positive or negative, every thought carries a frequency that influences how we "tune" to reality.

As discussed in Chapter 4, two people can experience the exact same situation yet respond in completely different ways. One sees opportunity, the other sees defeat. Why? Part of this difference lies in early childhood programming, the mental "software" written during our formative years. But there's more: the brain's integrity, its ability to process information and regulate thought patterns, is also shaped by the inputs it receives.

The Foundation of Positive Thinking

The brain doesn't operate in isolation. It is an energy-hungry organ, consuming nearly 20% of the body's fuel. Its performance depends on mitochondrial health, circadian rhythm, and environmental signals like light and electromagnetism. When the brain receives "junk inputs" such as artificial light, nnEMFs, toxic foods, or chronic stress, its neurons are more likely to misfire, producing distorted or chaotic thought patterns. But when it receives the right inputs, the electrical signalling becomes more coherent. Neurons fire in harmony, like a finely tuned orchestra, allowing clarity, focus, and better execution of free will.

This is the true foundation of positive thinking. It's not about forcing a happy thought in the middle of a storm. It's about creating the right environment for your brain to naturally default to clarity and resilience. For example, morning sunlight not only regulates circadian rhythm but also boosts serotonin, a neurotransmitter that elevates mood and makes positive thoughts more accessible.

When I was going through postnatal depression, positive thinking felt nearly impossible. My brain was misfiring under the weight of exhaustion, poor light exposure, and stress. No amount of affirmations could override that imbalance. But now that I live in sync with nature, positive thinking has become my second nature, a default setting rather than a struggle. It feels as though I've become the master of free will, able to tune it consciously, instead of being unconsciously tuned by it.

The Limitation of Thoughts & Beliefs

I once watched an interview where Eckhart Tolle was asked what he considered his greatest achievement.

His answer was simple: "I don't think much nowadays."

The audience laughed when he added, "But that's generally not well accepted by modern society."

I didn't laugh. Instead, I felt a deep sense of resonance.

The Innate Tuning Forks of Reality

We often glorify thinking, as if the more we analyse, calculate, and plan, the more control we'll have over life. But thoughts, whether positive or negative, can become like a pendulum: one moment swinging toward hope and excitement, the next toward fear and despair. Neither swing is stable. No matter how high the arc, the pendulum always falls back. That constant back-and-forth creates resistance, scattering our energy and adding disorder to the system.

From the lens of quantum possibility, this is why thoughts and beliefs are inherently limiting. They collapse the infinite field of potential into a narrow slice of reality based on our conditioned interpretations. Every belief acts like a filter, useful for navigating daily life, but also restricting what we can perceive or allow. Even positive beliefs can box us in, because they define what should be, rather than leaving space for what could be.

Modern neuroscience echoes this. The brain's default mode network, the region responsible for self-referential thinking and mental chatter, consumes enormous energy when left unchecked. It's why we can feel mentally exhausted without doing any physical work. But when we quiet this constant stream of thought, the brain shifts into coherence, opening space for presence, creativity, and insight.

In the present moment, there are no beliefs to defend and no thoughts to chase. There is only connection to life, to energy, to pure awareness. From that space, free will is not about endlessly choosing between "good" or "bad" thoughts, but about accessing the point where the pendulum is close to rest, where all possibilities remain open.

ATOMIC GLOW

The Power of the Empty Mind

This is how children live their lives, completely absorbed in the present moment, paying full attention to what's in front of them, free from the constant chatter of thought. That was how I lived in my childhood. If you recall the first two chapters, I described how I often acted without thinking, and somehow, things unfolded in the most miraculous way.

It's also how this book came into being. For more than 10 months, I have been wrestling with ideas, structuring outlines, and creating chapter names. But none of it came with ease and clarity. So eventually, I let it all go. I forgot the outlines, the structure, the "shoulds." One day, when I sat down at the computer, my hands simply began typing, no plan, no agenda. The words just flowed. Chapters shaped themselves, one after another, as if the book were writing me through my hands rather than the other way around.

This reminded me of my childhood heroes, the Shaolin monks and Bruce Lee. Shaolin Gong Fu is not magic. It is the art of emptying the mind and merging fully with the task at hand. When a monk strikes a rock, he first becomes one with it, sensing its strength, its weaknesses, its energy. When facing an opponent, he dissolves the boundary between himself and the other, attuning to their movements until he knows them from within. Out of this deep connection, the impossible becomes possible.

Bruce Lee captured this truth perfectly when he said, "Be water, my friend."

Water takes the shape of whatever holds it. In a cup, it becomes the cup; in a bottle, it becomes the bottle. Water can flow gently, yet it can also crash with unstoppable force. That is the power of the empty mind, fluid, adaptable, and unshakably present.

The Innate Tuning Forks of Reality

Around the world, countless miracles testify to this state of alignment. A mother lifting a car to save her trapped child. A man igniting paper using his fingertips with nothing more than focused intention. Athletes performing feats of endurance that defy all logic. Again and again, we see that when the mind is emptied of noise and aligned with nature's deeper design, an extraordinary intelligence flows through us.

This is not magic. It is the natural power we all carry, the power of the empty mind.

The Still Mind & the Creative Force

The mind is powerful when it thinks, but it is even more powerful when it rests in stillness, when it slips into the state of no-thought. In this state, the chatter of reaction falls away, and what remains is pure awareness.

Awareness is not bound by memory or fear. It does not judge, compare, or predict. It simply is. From this field of stillness, reality is not filtered through the limitations of past experiences or conditioned beliefs. Instead, creation arises directly from the unknown, from the infinite possibility that exists beyond the boundaries of thought. Here, the mind does not just imagine reality; it receives reality in its most unfiltered, limitless form. This is the creative force at its highest expression: creation from pure potential, not reaction to circumstance.

The Tuners of the Mind

The states of the mind are not only shaped by the mitochondria in the brain cells that operate on the quality of the inputs, such as light and electromagnetism, but also tuned, by design, by two other powerful "brains" within us: the gut and the heart. Both the gut and the heart have dense neural networks, biochemistry, and direct communication pathways with the brain. Together, they act like tuning forks, influencing what the mind thinks (or doesn't) and perceives, and ultimately creates reality.

The Truthful Gut

Our gut has its own nervous system, constantly talking with the brain through the vagus nerve. What's fascinating is that about 80–90% of the fibres in this nerve send information from the gut to the brain, not the other way around. That means the brain listens far more than it speaks.

These gut-born signals shape mood, cognition, stress, and even decision-making, carried through powerful neurotransmitters like serotonin, dopamine, and GABA. The gut microbiome not only shapes these signals but also provides the brain with essential short-chain fatty acids, the very fuel the brain needs to think, feel, and function clearly.

In short, the gut doesn't just digest food. It instructs, feeds, and tunes the brain. And when the gut falls out of tune, the mind quickly vibrates out of harmony.

Instinct and Intuition

Long before humans developed philosophy, complex language, or even a thinking brain, survival depended on the gut. It's often called our "first brain" for a reason. This primal compass guided our ancestors—warning them when food was safe or spoiled, when a path was dangerous, or when a person could be trusted.

The Innate Tuning Forks of Reality

This knowing doesn't arrive through words or logic. It speaks through sensation: a tightening in your belly, butterflies of excitement, nausea that won't go away, or an undeniable gut pull that says yes or no. While the mind spins stories, the gut deals in truth, raw, immediate, and unfiltered.

One story captures this perfectly. A man in the U.S. once described how he entered a busy venue, ready to meet friends. But the moment he stepped inside, his gut clenched so hard that he couldn't ignore it. Something felt deeply wrong. Against all logic, he turned around and left. Later that day, a mass shooting erupted in that very place. His gut had sounded the alarm long before his mind could have reasoned it out, and it saved his life.

The gut is the body's original compass, shaping not just what we think but whether we need to think at all. Some of the most powerful decisions aren't made by weighing endless pros and cons; they're born from a flash of certainty that bypasses the chatter of the mind.

This is instinct. This is intuition. An ancient wisdom that still lives in our bodies, guiding us through modern chaos. They act like a tuning fork, shifting us from overthinking into clarity. When the gut speaks and the mind listens, thought dissolves into presence, and presence is where new realities are born.

Your gut is proof that wisdom isn't only "out there" in books, theories, or experts. It's already within you, waiting to be connected and trusted.

The Buried Original Compass

For most of human history, we lived with this original compass. It was our ancient guide, keeping our bodies resilient and our instincts sharp.

At the heart of this compass lies the gut microbiome, a vast colony of trillions of microbes that depends on nature's rhythms to thrive, sunlight and darkness, the cycles of the seasons, and the grounding forces of the Earth. For millennia, this ecosystem flourished.

Today, that harmony has been disrupted by artificial light, nnEMFs, chemical overexposure, including relentless sanitation, and a lifestyle that's hurried, stressed, and overstimulated. As a result, the once-thriving colony of microbes has been eroded, in many cases severely diminished. The natural signals that were once clear and precise have now been muted, distorted, and drowned out by the noise of artificial living.

Over time, the gut's tuning power has weakened. Instinct feels muffled. Intuition seems unreliable. We second-guess instead of simply knowing. The ancient compass that once guided us through life's uncertainties becomes harder to read, replaced by external maps, apps, opinions, and endless noise from the outside world.

But the compass is not gone. It has only been buried. When we return to the rhythms of nature, the gut awakens again. Its voice becomes clear, steady, and trustworthy. And once more, it can tune the mind, guiding thought and stillness alike with the quiet authority of ancient wisdom.

The Counterintuitive Heart

Science has revealed astonishing truths about the heart, truths that confirm what wisdom traditions have whispered for centuries.

The Innate Tuning Forks of Reality

The Centre of Magnet

Often called our "third brain," the heart emits the strongest electromagnetic field in the human body, about 5,000 times stronger than the brain's. It contains more than 40,000 neurons, giving it its own form of memory and intelligence. Remarkably, it sends more signals to the brain than it receives, shaping our thoughts, emotions, and even the way we perceive the world. The heart is also a hormone factory, releasing powerful chemicals that regulate emotions, guide decision-making, and even spark creativity.

At its core, it is a relentless engine of life: with the highest density of mitochondria of any organ, the heart never takes a break, from its very first beat in the womb until our final breath.

In fact, before the brain has even formed, the heart is already alive, establishing the very first rhythm of existence. That primal pulse becomes the conductor of the body's symphony, orchestrating every system and setting the tone for our entire journey.

Think about babies: long before they can form words or logical thoughts, they communicate effortlessly through presence and feeling. Their hearts seem to speak directly to ours. This may explain why feelings born in the heart often travel beyond language, sometimes even beyond space, as though transmitted telepathically.

No wonder nearly every spiritual and wisdom tradition across history has placed the heart at the centre of being. Science now shows us why: the frequencies of our heart, whether calm, chaotic, loving, or fearful, tune the state of our mind, ripple through our nervous system, and shape the reality we experience.

ATOMIC GLOW

Frequencies of the Heart

The heart is far more than a physical pump; it is an energetic powerhouse that broadcasts our inner world outward. Every emotion we experience carries a frequency, and that frequency radiates through the heart's electromagnetic field, touching not just every cell within us but also the people and spaces around us.

In his Map of Consciousness, Dr. David Hawkins offers a scale to measure these states of being. At the bottom, guilt and shame vibrate at the lowest levels. As we move upward through courage and acceptance, the frequencies rise. Love resonates at 500, joy at 540, and enlightenment, the highest state, vibrates at 700 and above.

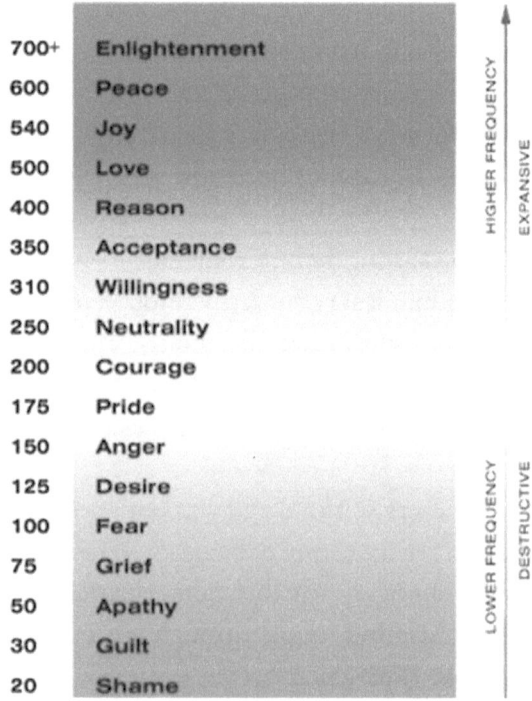

Graph 4: Vibration Chart

The Innate Tuning Forks of Reality

When your heart is open and filled with high-frequency states like love, gratitude, or joy, its field expands, becoming smooth, coherent, and magnetic. This coherence ripples outward, travelling at the speed of light, synchronising the cells of your body while connecting you with others. People sense it instantly. They feel drawn to your presence without needing words, as if your energy is quietly inviting them into harmony.

By contrast, when stress, anger, or fear take over, the heart contracts. Its electromagnetic field shrinks, becoming jagged and chaotic. Energy scatters rather than flows, and this incoherence is palpable. You've likely walked into a room and felt tension hang heavy in the air before anyone spoke; that was the heart's field broadcasting dissonance.

Your heart is always tuning reality, expanding or contracting, radiating coherence or chaos. And the beautiful truth is that, in every moment, you have the choice to shift its frequency.

The Power of the Open Heart

The heart is designed to open. Even after contraction, its natural state is expansion. It's as if the heart is always leaning toward harmony, waiting for us to choose alignment.

And that's the key: choice. At any moment, no matter the circumstance, we can choose to open or close the heart.

When my marriage was collapsing, everything in me wanted to contract, to fight, to blame. Yet in that moment, I counterintuitively opened my heart and let love in. To my surprise, what seemed broken began to heal. That single choice shifted the trajectory of my marriage and turned despair into possibility.

ATOMIC GLOW

Since then, I've made it a practice. Every time I face circumstances that naturally trigger contraction, anger, or fear, I pause and choose to open the heart instead. Time after time, life responds with something unexpected and beautiful. Miracles don't come from control; they come from coherence.

One such moment came on a quiet country road. Driving at 80 km/h, a pukeko suddenly flew across my path and was struck. I was devastated, watching it lie motionless. In that moment of heartbreak, I closed my eyes and radiated love toward the bird. To my astonishment, when I glanced back in the mirror after a while, I saw it slowly rise, pause, and eventually walk away. Friends called it a miracle. To me, it was the frequency of harmony in action.

I've also witnessed spontaneous healing in my own body. During meditation, while immersed in the amplified frequency of love, I felt a cooling, soothing sensation sweep across my brain. Afterward, a fever that had lingered for days vanished. At Dr. Joe Dispenza's week-long retreat, I watched people experience spontaneous healings as circles of participants generated fields of heart coherence, radiating love toward them. The energy was undeniable and transformational.

Stories like these are not rare. One of the most profound I've heard comes from Dr. Wayne Dyer's retelling of *Left to Tell* by Immaculée Ilibagiza, a survivor of the Rwandan genocide. Hidden for months, surrounded by violence, she shielded herself with the vibration of love and forgiveness. She said it was that frequency, an open heart, that protected her from being discovered and gave her strength to endure.

All of these experiences remind me that the heart's power extends far beyond romance or parental love. It is Big Love, a love that harmonises with all life, with nature, with the field itself. This is the ultimate tuning fork of human potential. When the heart is open, it not only transforms your inner state but reshapes the reality around you.

The Closed Modern Hearts

The power of an open heart is beyond what the mind can comprehend. Yet modern life has conditioned us to close our hearts out of fear, self-protection, or the illusion of control. We're taught that guarding our emotions makes us stronger, safer, and less vulnerable. But the truth is, a closed heart doesn't protect, it imprisons.

When the heart closes, energy stops flowing. Opportunities seem to slip away. Relationships feel distant. Life becomes repetitive, like walking in circles through the same old patterns. The body, too, pays the price: stress rises, the immune system weakens, and vitality fades. A closed heart is like a fortress, built to keep danger out but also keep love, joy, and infinite possibility from ever reaching in.

We've all felt it. Think of the person who's been hurt once and now trusts no one. Or the person who stays stuck in a job they hate, just to feel secure. Even those who have experienced the beauty of an open heart, like participants at Dr. Joe Dispenza's retreats, often find the feeling fades once they return to "normal life." I remember how stunned I was when I discovered many participants attended these retreats multiple times because the transformation didn't last. Without alignment in daily life, the walls rebuild, and the heart slips back into hiding.

Open Your Heart Naturally

The good news is that the heart doesn't want to stay closed. Its natural state is to expand, to connect, to open. When our mitochondria are nourished with the right inputs, the heart becomes easier to open. Openness shifts from being an effort into a second nature. When we resist, life itself nudges us, sometimes gently and sometimes dramatically, until we learn to open again.

We can also train the heart. Just as muscles strengthen with use, the heart's capacity to vibrate at higher frequencies grows with practice. Gratitude, compassion, and joy are not just fleeting emotions; they're states we can cultivate daily through practice.

And here's the beautiful cycle: when your heart opens, your mind begins to think more loving thoughts. Those thoughts lead to more loving actions. Those actions create more loving experiences. And the loop continues, expanding outward like ripples in a pond.

At its deepest level, an open heart is about aligning with the love of our greater Mother, nature herself. When you are in harmony with her rhythms, love arises naturally. You start seeing beauty everywhere, cherishing everything and everyone because you realise it's all connected. You're not separate from other people, from animals, or from the Earth; you are part of the same whole.

And when you live from that place, love is no longer something you "try" to feel. It becomes who you are.

CHAPTER 15

THE ART OF ATOMIC GLOW

"You are not a drop in the ocean.
You are the entire ocean, in a drop."
Rumi

Every story has a beginning, but the story of you did not start at birth. It began billions of years ago, in the silent dance of atoms, in the spark of life that first divided into two. Since then, life has been on an endless journey of evolution, expression, and expansion. From stardust to cells, from oceans to organisms, the same current of creation has flowed forward through time, culminating in you, in this very moment, reading these words, right here, right now. You are the living masterpiece of the universe's grand design, the most extraordinary creation ever made.

Evolution of Life Coded in You

I will never forget how the legendary Michael Singer once described the sheer miracle of human potential: the nine months we spend in the womb are like a condensed replay of the entire evolution of life.

The breathtaking story of life has unfolded across billions of years. It began with single-celled bacteria and archaea. Their union birthed eukaryotes, cells with mitochondria that transformed existence from mere survival into expansion. From there, life blossomed into ever greater complexity: fish swimming in the seas, reptiles roaming the land, mammals rising, and eventually, modern humans.

In just nine months, you retraced that same epic journey inside the sanctuary of the womb. From a single fertilised cell, you divided, multiplied, and specialised, forming tissues, organs, and systems with astonishing precision. For a moment, you resembled a tiny fish; later, reptilian; then, mammalian, until you emerged as a complete human being.

It's as if evolution compressed itself into a living symphony, replayed inside every mother's womb, so that each of us carries the whole story of life before our very first breath. This is not just poetry; it is biology reminding you of your origins.

Encoded in your DNA is the wisdom of billions of years, the proof that you are the culmination of life's greatest masterpiece, and the potential for so much more.

Endless Potential Surrounds You

And yet, your potential does not stop with your body. Beyond your skin, you radiate a biofield that resembles both the electron cloud of an atom and the vast electromagnetic force that binds the universe together. This biofield is your communication line, a living bridge between you and the quantum field: the invisible sea of energy that gives rise to all dimensions, all timelines, all galaxies, and even the so-called "empty space."

Scientists call it the unified field. Mystics call it the divine. Philosophers call it the eternal. Many simply call it God. Whatever the name, it is the womb of existence, the source from which all creation is born. It is a field of infinite possibilities, unlimited potential, where every version of reality that could ever exist is already present, waiting to be chosen, shaped, and experienced.

Nothing is separate from it. Not the stars above, not the ground beneath your feet, and not you. You are entangled with it, influenced by it, and, most astonishingly, capable of shaping it.

Many people have had glimpses of this truth. In deep meditation, when thoughts fade, the body dissolves, and only presence remains (what some describe as an out-of-body experience). In near-death experiences, when the physical body weakens, awareness expands into something vast, eternal, and timeless. In encounters with plant medicines, people often describe merging with an ocean of energy so boundless that the illusion of separation collapses.

In these states, the rules of time and space melt away. The walls of the 3D world dissolve. The illusion of "me" versus "everything else" falls apart, and what remains is oneness, the recognition that you are not apart from creation but creation itself. In that space, miracles are not only possible, they are natural. Reality bends, reconstructs, and responds to the frequency you bring.

Who You Truly Are

This is who you truly are: built from DNA that carries the entire evolutionary story of life within, and connected to the quantum field that holds the infinite story of possibility without. Every cell holds the wisdom of billions of years, while every heartbeat echoes the creative force that moves galaxies across the cosmos.

You are whole, both what can be seen and what remains unseen, both physical and spiritual, both the known and the yet-to-be-born. You are the living union of energy and matter, time and timelessness, form and formlessness.

You are invincible because the same intelligence that spins planets, births stars, and orchestrates the dance of the universe pulses through your veins and breathes through your lungs. Because your potential has no ceiling and your imagination knows no boundary.

It's written in your DNA. It shines through your biofield. The best part? You hold the keys to unlock it all.

Unlocking the Code Within

If you're already in awe of how mitochondria power your life, prepare to be even more amazed: they also play a decisive role in how your DNA is expressed. Your DNA carries the entire evolutionary history of life, but it's mitochondria that determine how much of that potential is brought to life. They do this through three vital outputs: biophotons, electromagnetic fields, and structured cellular water.

Biophotons act like a subtle Morse code, carrying light-based messages that help regulate which genes turn on and off. Think of your DNA as a piano: the code is all there, but mitochondria are the pianist, using light to choose which keys are played, and when.

Electromagnetic fields provide the cellular clock, ensuring that DNA responds with exquisite timing, so processes like repair, replication, and protein synthesis stay in perfect sync.

Structured water, in its gel-like, crystalline form, creates the stable medium that allows proteins to fold correctly, DNA to remain protected, and energy to flow efficiently.

When mitochondria are healthy, fed by natural sunlight, grounded in Earth's electromagnetic field, supported by oxygen, and nourished with electron-rich foods, they unlock more of the potential coded within your DNA. They are the keys that turn possibility into expression, so you don't just carry the story of life in your DNA, you live it fully.

Connecting to The Quantum Field

If mitochondria unlock the blueprint of life within, the heart unlocks your connection to the quantum field beyond.

The heart is a magnetic broadcasting station. Every time you choose to open it, your biofield expands. In that moment, you're not just influencing your own cells, you're connecting with the quantum field, exchanging information with the very source of creation.

This is why love and compassion feel so expansive. They are not just emotions; they are frequencies that bridge your body with the universe. When the heart opens, you enter into flow with life. You become magnetic. You become creative. You become limitless.

Love Your "Enemy"

One of the most common questions I hear is: "How can I love someone who is unkind, who hurts others, who feels like an enemy? They don't deserve my love. They deserve to be punished or taught a lesson." From global conflicts to personal relationships, we all encounter this resistance.

It's interesting how freely we give love to animals, to birds, to trees, yet when it comes to our own species, other humans, our hearts hesitate.

I've been there too. When a family member was unfair. When a colleague received the opportunity I thought I deserved it. Or when I looked at the chaos of wars, corruption, and hidden political agendas. My heart wanted to close, to judge, to resist.

Then I read something that changed everything: Neale Donald Walsch's *The Little Soul and the Sun.*

The Parable of the Little Soul

In the story, a little soul tells God that it wants to experience itself as the Light. God gently reminds the soul that it already is the Light.

But the little soul insists, "I want to feel it, to know it in my own experience."

To help, God explains that in order to experience what it means to be the Light, the soul will need contrast. For example, it cannot know forgiveness unless someone comes to give it an opportunity to forgive.

The little soul is excited, "I want to experience forgiveness."

Another soul then steps forward and says, "I will come into your life and do something so that you may forgive me. I may act in ways you call hurtful or cruel. In those moments, please remember who I am. I am sacrificing a part of myself so that you can experience who you truly are."

The little soul is astonished and asks, "Why would you do that?"

And the other replies, "Because I know who you are. You are the Light. And I love you."

A New Lens

The real gift of the story is this: those who challenge you are not here to break you, but to help you remember who you really are. Every hurt, betrayal, or injustice is an invitation to rise into your higher self.

When you shift your perspective, the "enemy" becomes a teacher in disguise. The unfair family member gives you a chance to practice compassion. The colleague who gets the opportunity you wanted reflects your own growth in patience and self-worth. Even the turmoil of global conflicts reminds us of the importance of unity and love.

No one is truly against you. Every person, even the difficult ones, plays a role in your evolution. They are not here to make your life worse; they are here to help you awaken to more of your potential.

When you truly grasp this, your heart softens. You begin to see that loving your "enemy" is not weakness; it is remembering the truth that every encounter is an opportunity to return to love, and every person is ultimately on your side.

ATOMIC GLOW

The Art of Atomic Glow

The universe is a mystery, 99% unseen, only a fraction understood. So too is our planet and ourselves. We are eternal students of life, destined to learn, unlearn, and rediscover. Perhaps we will never "know it all." And we are not meant to. That is the beauty of life: it is not a puzzle to be solved, but a symphony to be lived.

Living in Harmony

The art of living is about stepping into the flow of life, aligning with its rhythm, resonating with its energy, and connecting with its life forces, until you are no longer separate from life, but life itself.

Harmony begins within. It is the alignment of your three brains: the gut, the mind, and the heart. When they are in sync, your intuition speaks with clarity, your mind perceives with precision, and your heart remains open. It is the integration of body, mind, and spirit: the visible flesh and bone, the invisible realm of thought, and the eternal spark of your soul.

Yet harmony does not end with you. On the grandest scale, it is the union of you, Mother Earth, and the cosmos. You resonate with the sunlight and the electromagnetic pulse of the planet, and together you resonate with the forces that spin galaxies and ignite stars.

On the most intimate scale, harmony is the integrity of the atom: the nucleus at the centre, electrons and their cloud in constant motion, influencing and dancing with one another in coherence. That same atomic dance is happening within you.

Ultimately, harmony is the circle of life, the spiral that connects the vast cosmos to the tiniest particle. When you live by the laws of nature, you move in rhythm with the universe itself. Whatever comes is precisely what you need to fulfil your purpose, embody your potential, and leave your unique imprint on the fabric of existence.

Celebrating Imbalance

Perfect harmony does not exist. Taoism calls it Tao, the Middle Way, a path of balance and flow. In modern science, it's called equilibrium. But neither is meant to be a permanent state. They are not destinations to arrive at, but guides to help us navigate the journey.

Life is not designed to remain in perfect balance. If the heart ceased its rhythmic rise and fall, life would end. If the brain never responded to stimuli, you could not learn, adapt, or survive. Just as atoms vibrate and the universe oscillates, life itself depends on movement, rhythm, and change.

This is why we have a nervous system that responds to stress, and a sliver of conscious awareness that can choose. Life is meant to push us out of balance and challenge us. Stress, when temporary, strengthens; science calls this hormesis. Lifting weights tears muscle so it can grow stronger. Exposure to cold sharpens immunity and focus. Emotional challenges expand your capacity for compassion and courage.

The universe itself is powered by motion. Absolute stillness, perfect equilibrium, is not life but death, a return to the source from which we came. To live is to oscillate, to rise and fall, to expand and contract, to breathe in and out.

Our calling is not to escape this rhythm, but to embody it. As we grow, the swings soften, the centre steadies, and wisdom begins to flow with greater ease.

ATOMIC GLOW

Living Your Best Life Effortlessly

You are capable of climbing any mountain you set your heart and mind upon. Along the way, with your face pressed close to the rock, you may lose sight of the whole. You may stumble, take detours, or face dangers. Yet step by step, challenge by challenge, you will reach the peak. And that climb will shape you, teach you, and gift you resilience.

But there is another way: to live as if you are already at the summit.

From the mountaintop, the view is wide and limitless. You see not only beauty, but also the grand design, the hidden patterns, the simplest paths, and the obstacles you can step around with ease. From up there, you move with wisdom instead of willpower. You navigate without strain, and in that effortless state, life itself becomes miraculous.

Both paths hold wisdom: the climb builds strength; the summit offers freedom.

Living in harmony means choosing, again and again, to live from the summit. It is the conscious decision to take full responsibility for your destiny, to create an environment rooted in nature's healing forces that power your biology. It is walking each day with an open heart and a clear mind, no matter the circumstances.

Effortless living is always one choice away. By choosing harmony, you remember that life is happening for you and you are already its highest creation. From that knowing, you allow life to unfold with grace, release the weight of struggle with ease, and have the courage to let life carry you to your fullest potential.

This is the art of *Atomic Glow*: to move with the tides of change yet remain centred; to rise and fall with the breath of the universe while staying close to the Tao. Like Mother Nature herself, flowing with effortless grace yet holding immeasurable power. Commanding without force. Thriving without resistance.

And above all, always remember: you are the light. You are the glow the universe has been waiting to see.

AFTERWORD

OUR FUTURE—THE ERA OF THE HEART

As I close these chapters, my heart overflows with gratitude. Writing this book has been more than an act of sharing; it has been a wondrous journey of remembering. Every word and chapter that poured through me has reshaped and elevated me in ways beyond description.

Life reminds us again and again: we are not here to "master it all," but to keep growing through the cycle of learning, unlearning, and remembering.

One of the most profound realisations I had during this process was about something I have been resisting: the rise of Artificial Intelligence (AI). I often felt uneasy about how quickly it is advancing, replacing so much of the thinking we once relied on. I worried: if the brain is like a muscle, wouldn't outsourcing our thinking to machines weaken it over time?

But as this book took form, I found peace. AI is not the end of us; it is part of our evolution.

Humanity's story has always unfolded in phases. First, the **Era of the Gut**, when instinct and survival were our compass. Then, the **Era of the Brain**, when logic, reason, and invention shaped the modern world we live in today. Now, as AI takes on memory and calculation, it's the beginning of the **Era of the Heart**, an era where coherence, connection, and creation are our greatest powers.

By living in harmony with Mother Nature, your mind clears and your heart softens. Less thinking creates less noise, and with less noise, the old programs that no longer serve you begin to fade, leaving space for the subconscious mind—the cosmic library of possibilities— to take the lead. In this state, thought no longer arises from conditioned patterns but flows through you from the infinite field of possibility itself.

Your heart stays open, connecting you with the quantum field, the limitless potential. You create from the unknown, where everything is possible, instead of recycling the past or projecting a future constrained by what is already known.

As I picture my children and their children growing up in a rapidly evolving world, my heart is at peace. I no longer worry whether I'll leave the world better than I found it. I know the shift has already begun. The future is a better place, where generations to come will thrive in a more harmonious way of living, powered by nature, connected by heart.

About the Author

Alison Davis is a lover of life, a companion of nature, and a devoted truth seeker. Her path was forever changed when she was suddenly struck by severe postnatal depression and relentless insomnia, despite having lived in good health her entire life before. That unexpected descent became the turning point that awakened her natural curiosity and unstoppable determination to uncover the real root causes of vitality.

What began as a search to heal herself soon became a mission to help countless women facing the same hidden struggles. After exploring the depths of more than 500 books and immersing herself in the world of quantum physics and biology, Alison found herself guided back to life's simple, forgotten code—the original intelligence of nature that created and sustains us all.

This discovery not only restored her health but also transformed her family, home, and sense of purpose. It ignited a calling to share this wisdom with the world, so it is never forgotten again. That is why she wrote this book.

Today, Alison shares her message through her YouTube channel and the Atomic Glow podcast, where she inspires people worldwide to reclaim their vitality, awaken their inherent brilliance, and live fully alive. As a Certified Health Coach, she has also created signature video courses, designed for those seeking a more guided, hands-on path to transformation.

When she is not teaching or writing, Alison is often immersed in exploring wisdom and quantum science, whether through books or the living classroom of nature. You'll find her barefoot in her thriving food forests, strolling along the beaches of New Zealand, wandering through the woods, or simply soaking in the Sun, delighting in life's everyday miracles with her family and friends.

http://www.alisondavisglows.com

APPENDIX A

Mitochondrial Inputs & Outputs

This table provides a clear overview of the natural inputs that fuel your mitochondria, the outputs that power your life, and the distorted inputs of modern living. Use it as a quick reference guide whenever you want to reconnect with the essentials.

ATOMIC GLOW

Natural Inputs	Source	Natural Outputs	Distorted Inputs	Source of Artificial Inputs	Distorted Outputs
Sunlight	Full-spectrum sunlight (UV, visible, infrared) through the eyes and skin	Biophotons. ATP (energy). Structured cellular water. Electromagnetic field among cells and beyond the body	Lack of full-spectrum sunlight. Excessive artificial blue light	Screens. LEDs. Fluorescent bulbs	Overall, the distorted inputs lead to: Chaotic, incoherent biophoton emissions
Sunlight in Food	Electrons carried in natural, seasonal foods, where sunlight is captured and stored through photosynthesis	ATP (energy). Structured cellular water. Electromagnetic field among cells and beyond the body. Heat (infrared). Carbon dioxide	Fewer electrons. Excessive protons	Imbalanced electron/proton intake from chemically processed foods and products	Reduced or inefficient ATP energy. Insufficient, dysfunctional cellular water
Natural Electromagnetic Field	Earth's Schumann resonance and geomagnetic field	Electromagnetic field among cells and beyond the body	Lack of natural EMFs. Excessive non-native EMFs	Wi-Fi. Cell towers. Power lines. Dirty electricity. Satellites	Noisy, unstable cellular EMF signals. Excess or deficient heat output

MITOCHONDRIAL INPUTS & OUTPUTS

Natural Inputs	Source	Natural Outputs	Distorted Inputs	Source of Artificial Inputs	Distorted Outputs
Natural Water	Spring water. Mineral-rich water. Water from natural foods	Structured cellular water	Lack of natural water. Excessive contaminated water	Chemically treated foods and drinking water (e.g., fluoride, glyphosate, microplastics, heavy metals)	Dysregulated carbon dioxide levels
Oxygen	The air we breathe in	ATP (energy). Structured cellular water. Biophotons. Electromagnetic field among cells and beyond the body. Carbon Dioxide	Hypoxia (insufficient oxygen delivered to cells)	Sedentary, blue-lit lifestyle. Poor breathing habits. Air pollution. Deforestation	

REFERENCES

Mind, Heart, Spirituality

Achor, Shawn. The Happiness Advantage: The Seven Principles of Positive Psychology That Fuel Success and Performance at Work. Crown Business, 2010.

Dispenza, Joe. Breaking the Habit of Being Yourself: How to Lose Your Mind and Create a New One. Hay House, 2012.

Dispenza, Joe. Becoming Supernatural: How Common People Are Doing the Uncommon. Hay House, 2017.

Dispenza, Joe. Progressive Workshops and Week-Long Retreat (Australia) 2023.

Lipton, Bruce H. The Biology of Belief: Unleashing the Power of Consciousness, Matter & Miracles. Hay House, 2005.

Singer, Michael A. The Untethered Soul: The Journey Beyond Yourself. New Harbinger Publications, 2007.

Tolle, Eckhart. The Power of Now: A Guide to Spiritual Enlightenment. New World Library, 1997.

Tolle, Eckhart. A New Earth: Awakening to Your Life's Purpose. Penguin, 2005.

Walsch, Neale Donald. Conversations with God: An Uncommon Dialogue, Book 1-3. Putnam, 1995.

Walsch, Neale Donald. The Little Soul and the Sun: A Children's Parable Adapted from Conversations with God. Hampton Roads Publishing, 1998.

Katie, Byron. Loving What Is: Four Questions That Can Change Your Life. Harmony Books, 2002.

Taylor, Jill Bolte. Whole Brain Living: The Anatomy of Choice and the Four Characters That Drive Our Life. Avery, 2021.

Quantum Physics, Cellular Biology, EMF

Kruse, Jack. Selected Scientific Writings and Blogs. Published on JackKruse.com.

Kruse, Jack. Selected Scientific Blogs including the Consciousness Series (June–July 2025). Published on Patreon (member-only content).

DK. The Physics Book: Big Ideas Simply Explained. DK Publishing, 2019.

Pollack, Gerald. The Fourth Phase of Water: Beyond Solid, Liquid, and Vapor. Ebner and Sons, 2013.

Héroux, Paul. Building the Gulf of Opinions on the Health and Biological Effects of Electromagnetic Radiation. Frontiers in Public Health, July 2025.

Li, Y., & Héroux, P. Extra-low-frequency magnetic fields alter cancer cells through metabolic restriction. Electromagnetic Biology and Medicine, 33(4), 2013, pp. 264-275.

Wallace, Douglas C. A Mitochondrial Paradigm of Metabolic and Degenerative Diseases, Aging, and Cancer: A Dawn for Evolutionary Medicine. Annual Review of Genetics, 39, 2005, pp. 359–407.

Somlyai, Gabor. Defeating Cancer: The Biological Effect of Deuterium Depletion. Akadémiai Kiadó, 2001.

Popp, Fritz-Albert. Properties of Biophotons and Their Theoretical Implications. Indian Journal of Experimental Biology, Vol. 41, May 2003, pp. 391–402.

Van Wijk, R., & Van Wijk, E. P. A. An Introduction to Human Biophoton Emission. Forschende Komplementärmedizin und Klassische Naturheilkunde, 2005; 12(2): 77-83.

References

Van Wijk, E. P. A., Van Wijk, R., & Popp, F. A. Human Ultraweak Photon Emission and the Respiratory Chain. Journal of Integrative Biology, 2006; 10(2): 241–250.

Research, Data

World Health Organization. Cardiovascular Diseases (CVDs). 2023.

Alzheimer's Disease International (ADI) / World Health Organization. Dementia Statistics. 2023.

Wang, L., Wang, F-S., & Gershwin, M.E. Human Autoimmune Diseases: A Comprehensive Update. Journal of Autoimmunity. 2023.

International Diabetes Federation. Diabetes now affects one in 10 adults worldwide. IDF, 2021.

Lee, S. H., et al. Prevalence and Correlates of Brain Fog in the General Population: A Nationwide Cross-Sectional Study. Frontiers in Public Health, 2023.

Notes

ATOMIC GLOW

Notes

www.ingramcontent.com/pod-product-compliance
Lightning Source LLC
LaVergne TN
LVHW040050080526
838202LV00045B/3561